Cambridge Elements ≡

Elements in Grid Energy Storage
edited by
Babu Chalamala
Sandia National Laboratories
Vincent Sprenkle
Pacific Northwest National Laboratory
Imre Gyuk
US Department of Energy
Ralph D. Masiello
Quanta Technology
Raymond Byrne
Sandia National Laboratories

BEYOND LI-ION BATTERIES FOR GRID-SCALE ENERGY STORAGE

Garrett P. Wheeler
Brookhaven National Laboratory

Lei Wang
Brookhaven National Laboratory

Amy C. Marschilok
Brookhaven National Laboratory and Stony Brook University

CAMBRIDGE
UNIVERSITY PRESS

University Printing House, Cambridge CB2 8BS, United Kingdom

One Liberty Plaza, 20th Floor, New York, NY 10006, USA

477 Williamstown Road, Port Melbourne, VIC 3207, Australia

314–321, 3rd Floor, Plot 3, Splendor Forum, Jasola District Centre,
New Delhi – 110025, India

103 Penang Road, #05–06/07, Visioncrest Commercial, Singapore 238467

Cambridge University Press is part of the University of Cambridge.

It furthers the University's mission by disseminating knowledge in the pursuit of
education, learning, and research at the highest international levels of excellence.

www.cambridge.org
Information on this title: www.cambridge.org/9781009015707
DOI: 10.1017/9781009030359

© Cambridge University Press & Assessment 2022

First published 2022

A catalogue record for this publication is available from the British Library.

ISBN 978-1-009-01570-7 Paperback
ISSN 2634-9922 (online)
ISSN 2634-9914 (print)

Beyond Li-ion Batteries for Grid-Scale Energy Storage

Elements in Grid Energy Storage

DOI: 10.1017/9781009030359
First published online: June 2022

Garrett P. Wheeler
Brookhaven National Laboratory

Lei Wang
Brookhaven National Laboratory

Amy C. Marschilok
Brookhaven National Laboratory and Stony Brook University

Author for correspondence: Amy C. Marschilok, amarschilok@bnl.gov

Abstract: In order to improve resiliency of the grid and enable integration of renewable energy sources into the grid, the utilization of battery systems to store energy for later demand is of the utmost importance. The implementation of grid-scale electrical energy storage systems can aid in peak shaving and load leveling, voltage and frequency regulation, as well as emergency power supply. Although the predominant battery chemistry currently used is Li-ion, due to cost, safety, and sourcing concerns, incorporation of other battery technologies is of interest for expanding the breadth and depth of battery storage system installations. In this Element the authors discuss existing technologies beyond Li-ion battery storage chemistries that have seen grid-scale deployment, as well as several other promising battery technologies, and analyze their chemistry mechanisms, battery construction and design, and corresponding advantages and disadvantages.

Keywords: grid-scale energy storage, lead–acid battery, redox flow battery, sodium-ion battery, rechargeable aqueous zinc battery

ISBNs: 9781009015707 (PB), 9781009030359 (OC)
ISSNs: 2634-9922 (online), 2634-9914 (print)

Contents

1 Introduction

1.1 What Is Grid-Scale Battery Energy Storage?

The current energy storage technologies serve important functions pertaining to the generation, distribution, and other facets of the energy supply. A grid-scale battery energy storage system converts the energy collected from the grid or a power plant into a storable form and then discharges that energy at a later time to provide electricity or other grid services when needed [1]. Notably, in September 2013, the California public utilities commission passed a mandate for 1.3 GW of grid storage to be installed by 2020 [2]. In the 2018 New York State Energy Storage Order, the Commission adopted a statewide energy storage goal of installation of 1,500 MW by qualified energy storage systems by 2025, and 3,000 MW of qualified storage energy systems by 2030 [3]. Established grid-scale technologies, such as pumped hydro and compressed air energy storage, are capable of discharge times in tens of hours with large module sizes reaching 1,000 MW [4]. In contrast, various electrochemical batteries and flywheels are positioned around lower-power applications or those suitable for shorter discharge times (a few seconds to several hours). For example, to provide frequency regulation at Hazle Township, Pennsylvania, a 20 MW/5 MWh plant contains 200 flywheels and can source or sink 20 MW for 15 minutes. Sets of 10 flywheels share a containerized electronics module where each flywheel can store 25 kWh and has a design life of greater than 100,000 cycles [2].

1.2 Why Is Grid-Scale Battery Storage Advantageous for Renewable Energy Integration?

Currently, conventional energy storage technologies struggle to meet the demands of the grid. Most of the energy storage capacity currently on the grid is supplied by pumped hydroelectric power, compressed air energy, and mechanical flywheel systems, and each implementation has its disadvantages. Hydroelectric systems are the most widely utilized storage systems but are location-dependent, require a large capital investment, and suffer from low energy efficiency and environmental impact. Similarly, compressed air energy systems are also site-dependent and necessitate an underground reservoir to store the pressurized air using electric power during off-peak, which can be subsequently released during on-peak, driving the turbine/generator unit to produce electricity again [5]. While mechanical flywheels offer high power and efficiency, the use of flywheel accumulators is currently hampered by the danger of explosive shattering of the massive wheel due to overload [6]. As an on-demand supply system, grid-scale electrochemical energy storage systems (GSEESSs) serve fundamental roles in avoiding excessive power generation capacity to meet short-term peak energy needs. Some of these roles are the alleviation of the power supply gap of the grid, improving utilization efficiency of power generation equipment, avoiding the frequent start and stop cycle of the thermal power unit, decreasing the investment in power grid construction, and ensuring the safe and stable operation of the power grid system [1]. Furthermore, studies and observations around the world have demonstrated that interconnected power systems can safely and reliably integrate high levels of renewable energy from variable renewable energy (VRE) sources without new energy storage resources [7].

1.3 What Are the Key Characteristics of Grid-Scale Battery Storage Systems?

As a generalization, the battery technologies utilized in GSEESSs are expected to meet the following requirements: (1) peak shaving and load leveling; (2) voltage and frequency regulation; and (3) emergency power supply [1]. In order to meet these requirements, energy storage technologies that possess long storage duration, rapid response, long cycle life, low cost of the entire energy storage system, high power, and energy efficiency are needed [6].

1.4 Limitation of Li-Ion Batteries for Grid-Scale Battery Storage

The current markets for grid-scale battery storage in the United States and globally are dominated by lithium-ion chemistries [8]. Lithium-ion batteries (LIBs) represented more than 80% of the installed power and energy capacity of large-scale battery storage in operation in the United States at the end of 2016. Due to technological innovations and improved manufacturing capacity, lithium-ion chemistries have experienced a steep price decline over 70% from 2010 to 2016, and prices are projected to decline even further [7]. Li-ion batteries are characterized by a long cycle life (higher than 1,000 cycles), high efficiency (almost 100%), low self-discharge rate (2–8% per month), and wide operating temperature range (commercial Li-ion batteries may charge between 0 and 45° C and discharge between −40 and 65°C). Additionally, Li-ion batteries can be fabricated into a wide array of sizes and shapes, such as prismatic, spiral-wound cylindrical, and pouch cell designs in small (0.1 Ah) to large (160 Ah) sizes [9]. Despite the advantages mentioned here, there exist several major limitations and concerns for LIBs such as electrolyte flammability that necessitate research beyond Li-ion energy storage systems.

1.4.1 Lifetime Concerns

Volume expansion and shrinkage during Li-ion intercalation and de-intercalation often lead to detrimental structural changes, which result in performance loss upon cycling. In addition, the degradation of electrolyte as well as formation of irreversible solid-state electrolyte interphases (SEIs) often lead to deteriorated battery performance [10]. Along with this, prolonged exposure to heat reduces the battery lifetime, resulting in a pertinent issue in regions subject to high temperatures over prolonged periods, such as South Australia.

1.4.2 Safety Concerns

LIBs, being very energetic (particularly the lithium cobalt oxide cathode) and involving strongly exothermic processes, pose a risk of accidental short circuits subsequently leading to overheating and potential fire ignition that could extend to adjacent cells. Short circuits can potentially occur from various issues such as punctured separator membranes, either as a fault at fabrication (hence the crucial importance of rigorous production quality control) or as a result of erroneous or dangerous/unprotected operation [10].

1.4.3 Raw Material Sourcing

The likelihood of LIBs becoming ubiquitous for GSEESS is drastically reduced due to the fact that many battery components, including lithium and cobalt, are relatively scarce when compared to the demand on a global scale. Furthermore, these raw materials are mined in conflict zones, posing concerns regarding human rights and environmental provenance [10]. While the supply of lithium may be sufficient, as concerns of global shortages are speculative, the price of lithium will fluctuate drastically according to supply and demand, and diversification beyond Li-ion solutions can have benefit.

In this section, existing technologies beyond Li-ion battery storage chemistries (e.g., lead–acid batteries, flow batteries, nickel–cadmium [Ni–Cd] batteries, nickel–metal hydride [Ni–MH] batteries, sodium–sulfur [Na–S] batteries, and ZEBRA [Na-NiCl$_2$] batteries) that have seen grid-scale deployment, as well as several other promising battery technologies (metal-ion and metal–air batteries) for GSEESS are presented and analyzed in detail regarding their chemistry mechanisms, battery construction and design, and corresponding advantages and disadvantages.

1.5 Deployment Examples

The US Department of Energy's Office of Electricity houses a database of over 1,600 energy storage projects around the world that are either in operation, under construction, contracted, announced, offline, or decommissioned [11, 12]. The database is frequently updated and provides information about each project including the year of project announcement or completion, intended operational use, power, location, developer, and more. Insight into the distribution of grid-scale energy storage systems both in the process of development and current operation can be gleaned from this database. For the purposes of this section, only systems that store more than 100 kWh will be considered, and a breakdown of the different types of systems and stage of development is given in Table 1.1.

Electrochemical grid-scale energy storage is currently overwhelmingly dominated by the Li-ion system due to its high energy density and popularity. Figure 1.1 illustrates the number of deployment projects of different grid-scale energy storage systems, and what stage of development or development the project is in. Although Li-ion dominates the energy storage landscape, mature technologies like lead–acid and sodium batteries are still in use, and newer redox flow systems are gaining traction. Metal–air and Zn battery systems are emerging, and an increase in development of these types of systems is expected in the future. A few Ni-based systems are in use; however, the chemistry limitations of this mature technology may impede further deployment.

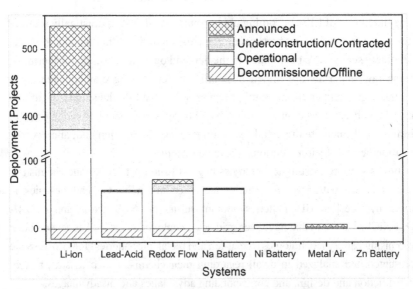

Figure 1.1 A visual representation of extent of deployment for each type of battery system and the stage of development of the various projects. Data collected on April 1, 2020 from the US DOE OE Global Energy Storage Database and the Distributed Energy Resources database for New York State [11–13].

Table 1.1 The stage of development and deployment of different electrochemical grid-scale energy storage systems. Data collected on April 1, 2020 from the US DOE OE Global Energy Storage Database and the Distributed Energy Resources database for New York State[11–13].

Type	Operational	Under Construction/ Contracted	Announced	Decommissioned/ Offline
Li Battery Systems				
Li-ion	381	52	101	15
Lead–Acid Systems				
Lead–Acid	55	0	2	12
Redox Flow Battery				
Vanadium	39	7	5	4
Zn bromide	12	2	0	7
Zn-ion	3	2	1	0
Zn–NiO	1	0	0	0
HBr	1	0	0	0
Nickel Battery				
Nickel–iron	3	0	0	0
Ni-Cd	2	1	0	0
Sodium Battery				
Sodium–sulfur	32	0	0	3
Na-NiCl$_2$	22	1	1	1

Table 1.1 (cont.)

Type	Operational	Under Construction/ Contracted	Announced	Decommissioned/ Offline
Na-ion	4	0	0	0
Metal–Air	0			
Zn–air		2	4	0
Zn battery				
Zn/MnO$_2$	1	0	0	0

With the rise of redox flow systems, a variety of different types of redox flow systems have been developed. Additionally, several different types of sodium batteries are currently in use. Figure 1.2 demonstrates the distribution of the current operational deployment of the various electrochemical grid-scale energy storage systems, and a break-down of the types of redox flow and sodium battery systems. Redox flow systems present the largest growth over recent years of all the beyond-lithium-ion systems. Currently, vanadium flow batteries make up nearly 70% of all operational redox flow battery systems. There are several other chemistry options available for redox flow batteries, which will be discussed further in a later section. The sodium batteries include high-temperature sodium–sulfur (NaS) and $Na/NiCl_2$ systems, with a few implementations of Na-ion battery systems. Due to the high temperature requirements of NaS and $Na/NiCl_2$ systems, increased future deployment of other low-temperature and room-temperature Na-ion batteries is anticipated.

When selecting the suitable energy storage systems, several key metrics need to be taken into consideration. Table 1.2 compares key properties of several commercially available battery systems, including lead–acid, lithium-ion, sodium–nickel chloride (NaNiCl), vanadium redox flow battery (VRFB), nickel–cadmium (NiCd), zinc–bromine flow battery (ZBFB), and sodium–sulfur (NaS) batteries [14].

2 Existing Battery Technologies for Grid-Scale Electrical Energy Storage

2.1 Aqueous Lead–Acid Batteries and Ni-Based Batteries

2.1.1 Lead–Acid Overview

Lead–acid batteries are one of the oldest types of rechargeable batteries, having been developed by Gaston Planté in 1859 [15, 16]. Through the use of lead and lead oxide electrodes in an aqueous sulfuric acid electrolyte, the lead–acid battery has been an integral part of the battery landscape for over 100 years. There are two main types of lead–acid batteries: starting, lighting, ignition (SLI) batteries and deep-cycle batteries. SLI batteries have thin electrodes for fast discharge of minimal capacity; SLI lead–acid batteries are used to start vehicles. Deep-cycle batteries have thick electrodes meant for extended duration of use and greater depths of discharge; deep-cycle lead–acid batteries are utilized for grid-scale energy storage. Lead–acid batteries have several advantageous properties for their use in grid energy

Table 1.2 Properties of the different types of energy storage systems [14].

Battery Type	Energy Density (Wh/L)	Power Density (W/L)	Nominal Voltage (V)	Life Cycle	Depth of Discharge (%)	Round-Trip Efficiency (%)	Estimated Cost (USD/kWh)
Lead-acid	50–80	10–400	2.0	1,500	50	82	105–475
NaS	140–300	140–180	2.08	5,000	100	80	263–735
NaNiCl	160–275	150–270	-	3,000	100	84	315–488
NiCd	60–150	80–600	1.3	2,500	85	83	-
VRFB	25–33	1–2	1.4	13,000	100	70	315–1,050
ZBFB	55–65	1–25	1.8	10,000	100	70	525–1,680
Li-ion	200–400	1500–10,000	4.3	10,000	95	96	200–1,260

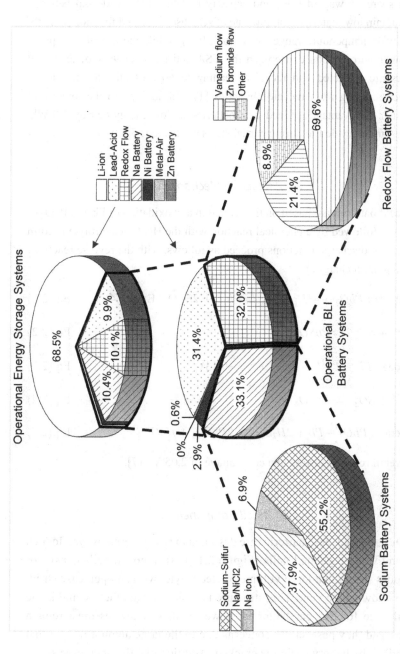

Figure 1.2 Pie charts representing the operational electrochemical energy storage systems. Included is the distribution of the types of sodium battery and redox flow battery systems. Data collected on April 1, 2020 from the US DOE OE Global Energy Storage Database and the Distributed Energy Resources database for New York State [11–13].

storage. The technology, production, and market presence are exceptionally mature; as of 2017, lead–acid batteries had the largest rechargeable battery market share by way of sales and capacity (MWh) [17]. Lead–acid batteries also maintain low material cost, low rate of self-discharge, and function properly over a wide temperature range (40°C to –30°C). While the use of lead poses a serious environmental concern, in the USA and EU over 99% of lead–acid batteries are collected and recycled, leading to 60% of all lead production each year coming from recycled batteries [17]. In addition to the toxicity of lead, other disadvantages of lead–acid batteries are low energy density (30 Wh/kg) and limited cycle life during partial states of charge (PSoC).

2.1.2 Reaction Mechanism

Lead–acid batteries are based on the reduction and oxidation of Pb and PbO_2 to form Pb^{2+} followed by a chemical reaction with the H_2SO_4 electrolyte to form $PbSO_4$. The discharge reactions proceed as follows, with the reverse reactions occurring upon charging.

$$\text{Cathode}: PbO_2 + 4H^+ + 2e^- \leftrightharpoons Pb^{2+} + 2H_2O \quad \text{E}° = 1.46\text{V} \qquad \text{Eq.}(2.1)$$

$$Pb^{2+} + SO_4^{2-} \leftrightharpoons PbSO_4 \qquad \text{Eq.}(2.2)$$

$$\text{Anode}: Pb \leftrightharpoons Pb^{2+} + 2e^- \quad \text{E}° = -0.13\text{V} \qquad \text{Eq.}(2.3)$$

$$Pb^{2+} + SO_4^{2-} \leftrightharpoons PbSO_4 \qquad \text{Eq.}(2.4)$$

$$\text{Overall}: PbO_2 + Pb + 2H_2SO_4 \leftrightharpoons 2PbSO_4 + H_2O \qquad \text{Eq.}(2.5)$$

The overall reaction results in a cell voltage of 2.05 V [17].

2.1.3 Cell Components

Cell Design Lead–acid batteries are designed in two different ways: flooded cells and valve-regulated cells (Figure 2.1) [18]. Flooded cells consist of electrodes sitting in a large volume of electrolyte, with an open case structure to allow gas to vent. This structure is easier to manufacture and is the typical style for SLI car batteries; however, these batteries must remain upright, and they pose safety concerns due to the large amount of acid that can spill if the battery casing is cracked. Additionally, the open case structure of flooded cells causes substantial loss of electrolyte over time due to water electrolysis, which requires battery maintenance to replenish the lost

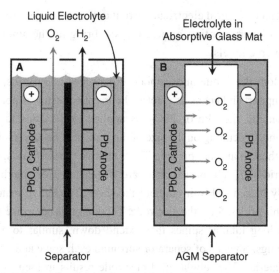

Figure 2.1 Schematic of a (a) flooded lead–acid battery and a (b) valve-regulated lead–acid (VRLA) battery.

water. Valve-regulated lead–acid (VRLA) batteries are sealed and use a gel electrolyte formed from H_2SO_4 with silica particles or electrolyte permeating a silica fiber separator called an absorbent glass mat (AGM). For VRLA cells with an AGM separator, a compression force is applied to the electrodes to hold them in intimate contact with the electrolyte-containing separator. Water electrolysis can still occur; however, generated oxygen can be recombined at the anode to re-form water; see Eq. (2.6) [19].

$$Pb + \frac{1}{2}O_2 + H_2SO_4 \rightarrow PbSO_4 + H_2O \qquad \text{Eq.}(2.6)$$

The generated hydrogen, however, does not re-form water due to kinetic limitations, and therefore VRLAs require a one-way valve to vent the pressure buildup from hydrogen in the system, which is how VRLAs get their name. Due to the lower maintenance requirements for VRLAs, this cell design is the predominant cell design used for grid-scale energy storage.

Electrolyte In lead–acid batteries, a highly concentrated aqueous solution of ~35% sulfuric acid is typically used for the electrolyte. Several challenges are associated with the aqueous electrolyte. Firstly, the reduction and oxidation of water to hydrogen and oxygen gas over time will decrease the electrolyte volume and can lead to drying out of flooded type cells and requires a pressure release valve in VRLAs [17]. Typically, by controlling charging voltage, water electrolysis can be minimized. Additionally, proper selection of electrode grid electrical current collector alloys can allow for an increase in the overpotential for electrolysis, bringing the amount of water loss to reasonable levels.

Electrodes Both the anode and cathode utilize lead alloy grids, which give mechanical support and act as an electrical current collector. A porous paste of the active material (i.e., Pb or PbO_2) is applied to the grids to form plates (electrodes) and, depending on the size of the lead–acid battery, multiple sets of anode and cathode plates are present. There are two types of grids: pasted plate and tubular grids [17]. The pasted plates are, in general, a rectangular sheet with systematically placed holes throughout the sheet. The active material is then pasted onto this grid to form the electrode. Tubular grids comprise a crossbar at the top with long tubular spines that extend down, similar to a comb with elongated prongs. Sheaths of separator surround each spine and are filled with the active material. The tubular grid electrode results in long cycle lives by containing the active material and holding it in close contact with the electrical current collector grid.

The alloy used in the cathode grid depends on the type of lead–acid cell being used [15, 20]. Flooded cells typically use a lead–antimony alloy, because antimony incorporation aids in deep-discharging cyclability. However, antimony lowers the water reduction overpotential and causes an increase in the amount of hydrogen produced. For a flooded cell this is not critical, as water needs to be added periodically anyway; but for VRLAs, lead–antimony alloy grids cannot be used due to increased electrolyte loss. A lead–calcium–tin or pure lead grid is typically used in VRLA cells. Tubular grids are very advantageous for flooded cells as they hold the active material close to the grids; however, for VRLAs, adequate compression of tubular grids with the AGM separator is challenging. For this reason, commercial VRLAs use only pasted plate grids.

Lead–acid anodes always use pasted plate grid current collectors and utilize either low-antimony alloys or lead–calcium–tin alloys [15, 20]. More recently, changes in the anode active materials have caused a pivotal change in lead–acid technology, bringing about the term advanced lead–acid battery. The incorporation of conductive carbon with lead in the anode paste or the use of a high-surface-area carbon anode can drastically improve the rate performance of the battery [16–18, 20]. The carbon acts in two main ways: it helps inhibit excessive buildup of $PbSO_4$ (this will be discussed later), and it leads to capacitive effects, giving the battery supercapacitor-like properties. This is extremely advantageous when pairing battery systems with renewable energy, as the rapid fluctuations in the renewable energy input can be accommodated by changes in the batteries' PSoC [16]. A dual anode can even be used, a conventional pasted plate electrode and a lead–carbon electrode; this design architecture has been designated as an Ultrabattery [21]. The Ultrabattery improves upon both the standard lead–acid battery and the supercapacitor-style advanced lead–acid battery by improving stability during PSoC, improving the high rate of charge acceptance, and maintaining the energy density and voltage profile of a standard lead–acid battery [17, 18]. The advanced lead–acid battery and Ultrabattery make up the majority of the recently deployed or announced lead–acid grid-scale energy storage projects [11, 13].

Separators For flooded cells, a wide variety of materials can act as a separator: for example, microporous polyethylene, polyvinyl chloride, or rubber [17]. These materials only need to be porous enough to allow for good electrolyte mobility and strong enough to resist penetration by lead growth. For VRLAs, typically the AGM separator is used [17]. This is made from glass microfibers, which can separate the electrodes while absorbing the H_2SO_4 electrolyte. A key property of the AGM separator, in addition to absorbing electrolyte, is that it also maintains channels of void space. This void space is critical for allowing transport of oxygen gas from one electrode to the other to enable oxygen reduction back to water as described in Eq. (2.6).

2.1.4 Current Status and Practical Challenges

Due to the low cost and maturity of the technology, there is an extensive deployment of lead–acid batteries around the world [11, 13]. However, many of the larger deployment facilities have been decommissioned and replaced with newer Li-ion technologies or dynamic volt-ampere reactive (D-VAR) systems [11]. One existing large-scale deployment example is the 2013 Notrees Windpower project that has 24 MWh of advanced lead–acid battery storage capacity [16]. The 153 MW wind farm utilizes VRLA lead–acid batteries to manage and stabilize energy flow and enhance electrical output during nonpeak electrical generation.

Several practical challenges face the lead–acid battery, which are responsible for the phasing out of grid-scale lead–acid energy storage in favor of other technologies. For integration with renewable energy, a constant PSoC is present during the fluctuations of renewable energy input [17, 18]. This can lead to irreversible sulfation of the Pb anode. On a typical discharge, $PbSO_4$ is generated, but as very small crystallites all over the surface. Irreversible sulfation occurs when these crystals grow significantly in size to the point that electrical connectivity is lost, or the crystals cannot be fully dissolved upon charging. This is one of the leading causes of failure in a lead–acid battery [17, 18]. Due to the issue of sulfation during PSoC operation, cycle lives for lead–acid batteries are relatively low, making them less economical for grid-scale energy storage applications. Additionally, relatively low rates of charging and discharging are observed for lead–acid batteries compared to other technologies. When considering the low energy density of lead–acid batteries, it is apparent that new energy storage technologies are superior to that of lead–acid batteries, and the continual phaseout due to the preference for newer technology is likely to continue.

2.1.5 Aqueous Ni-Based Batteries

Similar to lead–acid batteries, nickel–iron (NiFe), nickel–zinc (NiZn), nickel–cadmium (NiCd), and nickel–metal hydride (NiMH) batteries were discovered, optimized, and commercialized decades ago. All four systems utilize the same $Ni(OH)_2$-to-$NiOOH$ reaction at the cathode, and a similar highly alkaline aqueous electrolyte. For the NiFe system, upon discharge, Fe metal is oxidized to $Fe(OH)_2$ in solution, generating a discharge voltage of ~1.05 V [15, 16]. For the NiZn system, upon discharge, Zn metal is oxidized to $Zn(OH)_4^{2-}$ in solution, generating a discharge voltage of ~1.75 V, which is the highest voltage of all the aqueous Ni systems [22]. For the NiCd system, upon discharge, Cd metal is oxidized to $Cd(OH)_2$ in solution, generating a discharge voltage of ~1.2 V [23, 24]. For the NiMH system, upon discharge, the MH anode is reduced to a pure M anode and H^+ is released into the solution, generating a similar discharge voltage of ~1.2 V [25]. All three systems, NiFe, NiCd, and NiMH, have higher specific energies than that of lead–acid batteries at approximately 30–50 Wh/kg, 50 Wh/kg, and 100 Wh/kg respectively [15, 20, 25].

The NiFe system has the lowest specific energy; but due to its inexpensive materials and structure, benign environmental impact, and long cycle life, NiFe is the most promising of the Ni-based battery systems for grid-scale energy storage. Iron Edison is a producer of high-capacity stationary NiFe batteries and has deployed several 100 to 500 kWh installments around the United States [11, 16]. NiZn is desirable due to its high voltage, low component cost, and high recyclability [22, 26]. The main drawback for NiZn is limited cycle life due to Zn metal dendritic growth penetrating the separator and shorting the cell. As a result there are no large-scale deployments of this battery system; however, improvements to the Zn anode are being heavily researched and will be discussed in more detail in the Section 3.3. NiCd has a slight improvement in specific energy; however, the cadmium anode gives rise to concerns about toxicity and environmental impact of large-scale utilization of the NiCd system. Additionally, NiCd batteries are well known to possess a "memory" effect relating to decreased capacity upon charging a non–fully discharged battery [23]. This memory effect severely limits the application of NiCd grid-scale batteries, as full discharging is required before the batteries can be charged again. NiMH batteries improve upon the design of the NiCd system by removing the toxic Cd anode, eliminating the memory effect, and increasing specific energy. As such, NiMH batteries were utilized for the early hybrid and electric vehicles in the 1990s and early 2000s [23, 25]. Unfortunately, although the NiMH battery is superior in specific energy to NiFe and NiCd systems, the increased cost to produce these batteries has kept them from being used in any grid-scale applications. In general, the Ni-based batteries are mature battery technologies; however, they still require further improvements to constitute incorporation into the grid energy storage landscape.

2.2 Redox Flow Batteries (RFBs)

2.2.1 Overview

Redox flow batteries (RFBs) have their origins in the 1960s and use two circulating soluble redox couples as electroactive species to store or deliver energy [27]. Liquid electrolytes are pumped from storage tanks to flow over electrodes, where chemical energy is converted to electrical energy during discharge and the reverse process happens during charge. The electrolytes flowing through the cathode current collector and anode current collector are referred to as catholyte and anolyte, respectively. The redox-active species undergo oxidation or reduction reactions when they are in contact or close proximity to the current collectors during operation. An ion-selective membrane separates the positive and negative redox compartments and selectively allows cross-transport of nonactive species (e.g., H^+, Na^+, Cl^-) to maintain electrical neutrality and electrolyte balance [28].

Redox flow batteries possess several advantages. (i) Different from traditional batteries, power and energy are decoupled in RFBs; through separation of design, modular and flexible operation is provided to the system. Specifically, the power of the RFB is determined by the size of the electrodes and the number of cell stacks, whereas the energy density is linked to the concentration and volume of electrolytes. (ii) Depending on the application, the RFB could conceivably be adjusted for storage over variable time scales from a few hours to days, which is another important advantage for renewable integration. (iii) Facile cell design allows for construction of large-scale systems based on module design. Despite the numerous advantages of RFBs, several limitations still exist. (i) The system has many requirements for pumps, sensors, reservoirs, and flow management [29]. (ii) There is still a need for designing cost-effective membranes that can control long-term ion crossover effects. (iii) The concentration of redox couple is limited to ~8 M, which limits the energy density of RFBs to about 25 Wh/kg [27]. Lower-cost redox couples with high solubility remain essential for this technology to succeed. To further increase the energy density, a semisolid flow battery design has been introduced [30], which contains conducting inks (i.e., $LiCoO_2$ and $Li_4Ti_5O_{12}$ in nonaqueous electrolyte solutions) rather than solids. The redox-active suspension circumvents the problem of low solubility of the metal ion redox couples in aqueous solutions by increasing the concentration of active species to the 10–40 M range, which is at least five times higher than traditional redox flow systems. Since these batteries function based on Li-ion battery chemistry, their details will not be covered in this section.

Several types of RFBs have been reported, based on different anolyte and catholyte chemistries: all vanadium redox flow batteries (VRBs), zinc/bromine flow batteries (ZBB), polysulfide/bromine flow batteries (PSBs), iron/chromium flow batteries (ICB), vanadium/cerium flow batteries, and others. Among the RFBs, ZBB, VRB, PSB, and ICB have been demonstrated at levels of a few hundred kW and even multiple MW [28]. In this section, the most mature system VRB will be discussed in detail as model systems, including operational chemistries and cell components. That will be followed by a brief introduction of other RFBs, along with an overview of status and challenges of the technologies.

2.2.2 All-Vanadium RFBs

The all-vanadium RFBs are promising batteries for large-scale energy storage systems due to their scalability and flexibility, excellent durability, high round-trip efficiency, and little environmental impact [31–33].

2.2.3 Reaction Mechanism

VRBs exploit the capability of vanadium to exist in solution in four different oxidation states. By having only one cationic element, the crossover of vanadium ions through the membrane upon long-term cycling in VRBs is less deleterious than with other chemistries. The following reactions happen during the discharge process in VRB.

$$\text{Cathode}: VO_2^+ + 2H^+ + e^- \rightarrow VO^{2+} + H_2O \qquad \text{Eq.(2.7)}$$

$$\text{Anode}: V^{2+} - e^- \rightarrow V^{3+} \qquad \text{Eq.(2.8)}$$

Overall reaction : $VO_2^+ + 2H^+ + V^{2+} \rightarrow VO^{2+} + V^{3+} + H_2O$ Eq.(2.9)

The overall reaction results in a cell voltage of 1.26 V at 25°C [28].

2.2.4 Cell Components

Electrolytes In VRBs, the anolyte and catholyte are solutions containing V(III)/ V(II) and V(V)/V(IV) respectively. H_2SO_4 has been the commonly used supporting electrolyte in both the anolyte and catholyte (Figure 2.2). The solution of V(IV) ions is prepared by dissolving $VOSO_4$ in H_2SO_4 [34]. The solubility of $VOSO_4$ has been confirmed to decrease with increasing H_2SO_4 concentration, but increase with operating temperature [35]. Hence, the concentration of vanadium and total SO_4^{2-} is usually controlled at less than 2 and 5 M, respectively. V(IV) exist as blue oxovanadium ion $[VO(H_2O)_5]^{2+}$ in H_2SO_4, which is stable within the temperature range of 33 to 67°C. The sulfate anions are weakly bound to the vanadyl ions creating the second coordination sphere, which may impact the redox reactions of V(IV) by affecting the proton and water exchange kinetics of vanadyl ions. V(V) ions exist as yellow dioxovanadium ion VO_2^+

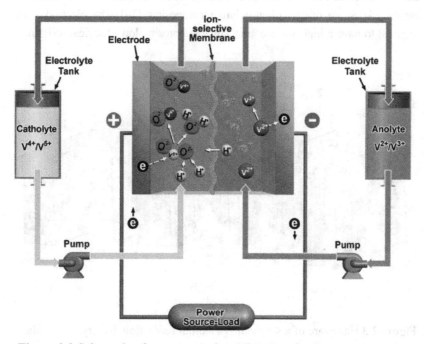

Figure 2.2 Schematic of components in a VRB. Reprinted with permission from [28]. Copyright (2011) American Chemical Society.

and its hydrated form of $[VO_2(H_2O)_4]^+$ or $[VO_2(H_2O)_3]^+$. The energy density of VRBs is enhanced with increasing concentration of vanadium species. However, the concentration of vanadium is limited by the solubility of the vanadium ions. Increasing the vanadium concentration to above 2 M in H_2 SO_4 will lead to the formation of solid V_2O_5 in V(V) solution at temperatures higher than 40°C and VO in V(II) and V(III) solutions lower than 10°C. Hence, it is critical to optimize the operating conditions, such as temperature, the concentration of vanadium and sulfuric acid, as well as the state of charge of the electrolyte to improve the stability of electrolytes. Generally speaking, the vanadium concentration in most practical VRBs is limited to under 2 M for the temperature range of 10–40°C.

Electrodes/Bipolar Plates In VRBs, electrodes are often integrated with bipolar plates into one component. Specifically, the bipolar plate serves as a current collector, while separating the anolyte and catholyte on each side. In addition, it functions as supports for the porous electrodes and directs the flow of the electrolytes by incorporating flow channels on both sides (Figure 2.3). The characteristics of bipolar plates include low bulk and contact resistance, structural and mechanical stability, as well as chemical compatibility with electrolytes during battery operations. Given the strong acidic conditions in VRBs, the materials of choice for bipolar plates are often carbon-polymer composites [36]. The electrodes are required to have a high surface area, suitable porosity, low electrical resistance,

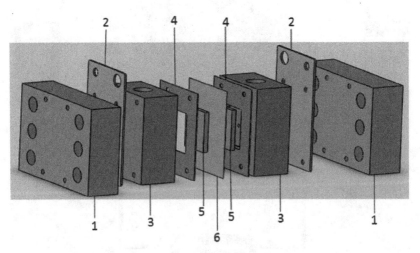

Figure 2.3 Hardware of a single all-vanadium redox flow battery. 1-end plate, 2-current collector, 3-bipolar plate, 4-gasket, 5-electrode, and 6-membrane. Reprinted with permission from [28]. Copyright (2011) American Chemical Society.

and high electrochemical activity. The limited choices of electrodes include carbon-based materials in forms such as felt or porous structures [36, 37].

The reaction in the positive half-cell involves oxygen transfer: (i) First, VO^{2+} ions migrate from the solution to the electrode surface, exchanging protons with phenolic groups on the electrode surface; (ii) second, electron transfer from VO^{2+} to the electrode through the C-O-V bond, which breaks one C-O bond and forms VO_2^+; (iii) the VO_2^+ exchanges with a proton from solution and diffuses back into the solution. Such reactions are the reverse of the charge process. In the negative half-cell, the reaction involves electron transfer: (i) first, upon charge, V^{3+} diffuses from the solution to the electrode surface and replaces the hydrogen in phenol groups; (ii) second, electron transfer happens from the electrode surface to the V^{3+} along the C-O-V bond to form V^{2+}; (iii) finally, the V^{2+} exchanges with protons and diffuses back to the solution. The discharge reactions are the reverse of the as-described charging process. To improve both coulombic and voltage efficiencies of VRB, increased concentrations of surface functional groups such as C=O and C-OOH are required, which are generally formed through acid treatment of the carbon electrodes. These functional groups not only increase the hydrophilicity of graphite felt but also serve as active sites for the electrochemical reactions.

Membranes and Separators Membranes in VRB play the critical role of separating the catholytes and anolytes compartments while allowing transport of charge carriers (i.e., H^+, SO_4^{2-}) to maintain the electrical neutrality. Several requirements are important to the membranes, which include (i) high ionic conductivity to reduce resistance and power loss, (ii) ion selectivity to minimize vanadium ion transport to reduce capacity and energy loss, (iii) limited water transport to maintain the balance of, and ease maintenance of catholyte and anolyte, (iv) high chemical stability toward the strong acidic and highly oxidative environment in the positive cell, (v) excellent mechanical and structural stability needed for the large size, and (vi) low cost to achieve commercialization.

Both cation and anion exchange membranes have been explored for VRB applications, among which the most widely studied are the Nafion membranes [38] that are highly conductive to protons and are chemically stable toward the strong acidic and oxidative environment. One great challenge in VRB is that the vanadium ions tend to transport from one side of the cell to the opposite side and react with other vanadium ions with different oxidation states, leading to loss of cell capacity and reduction of energy efficiency. The transport rate highly depends on the concentration of vanadium ions and sulfuric acid, the SOC of electrolyte, the thickness and pore size of membranes, and the operating temperature. In addition, the net water transfer often induces the cross-transport of negative and positive half-cell electrolyte solutions [28]. To improve the selectivity and minimize the cross-water and vanadium transport, varied approaches were taken to modify the membranes, such as Nafion-based hybrid membranes, which include other polymer components to reduce permeability of vanadium ions [39]; and inorganic-layer-coated Nafion^{TM} membranes [40] to reduce permeability of both vanadium ions and water. In addition to cation membranes, several anion exchange membranes have been reported in VRBs, including a sulfonated polyethlene membrane which allowed cross-transport of SO_4^{2-} and $VO_2SO_4^-$ as well as quaternized poly (phthalazinone ether sulfone ketone) (QPPESK), which demonstrated chemical stability in VO_2^+ solution [28].

2.2.5 Other RFBs

Several other systems have demonstrated their application potentials in RFBs [41]. Their standard cell potentials are depicted in Figure 2.4 [28]. Among these other RFBs that are not all vanadium, zinc/bromine flow batteries (ZBB), Br^-/Br_2 vs Zn^{2+}/Zn; polysulfide/bromine flow batteries (PSBs) with Br_2/Br^- vs S/S^{2-} redox couples; and iron/chromium flow batteries (ICB) with Fe^{3+}/Fe^{2+} versus Cr^{3+}/Cr^{2+} were also demonstrated at scales up to 100 kW and even MW levels. These demonstrated flow batteries are briefly reviewed in the following sections.

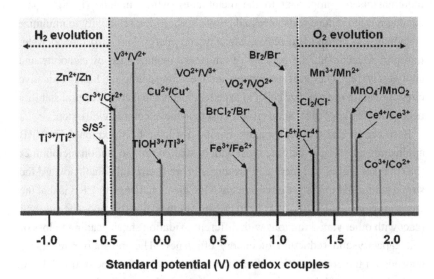

Figure 2.4 Redox potentials (vs. standard hydrogen electrode) of various redox couples. Reprinted with permission from [28]. Copyright (2011) American Chemical Society.

ZBB The cathodic reaction involves the reduction of Br_2 and oxidation of Br^-, while Zn strips and plates onto the anode. ZBB is considered to be a hybrid flow battery, in which one of the electrodes involves nonliquid reactants [42]. The cell reaction gives a standard voltage of 1.85 V [28]. At the positive electrode, Br^- ions are converted to Br_2 during charging and vice versa during discharging. Complexing agents are required to reduce the amount of freestanding Br_2, which is a serious health hazard. At the negative electrode, Zn is reversibly deposited from the ions. The synergy between the dissolved redox couple and the metal anode offers a great improvement in energy density due to the small volume of anolyte [42]. However, the power/energy relationship of a ZBB is less flexible than that of the VRB, as its available energy is limited by the Zn electrode area. ZBB offers higher reversibility, voltage, and energy density than those of VRB and PSB (Table 2.1). In addition, the abundant, low-cost chemicals that are employed in the system are of interest to large-scale deployment. However, Zn dendrite formation and reversibility of Zn electrodeposition remain drawbacks of these batteries.

PSB PSBs utilize Na_2S in the anolyte, coupled with NaBr as the catholyte, which are both abundant and soluble in aqueous solutions and are of reasonable cost. Upon charge, the bromide ions are oxidized to Br_2 and complexed as Br_3^- in the cathode side, while the S_4^{2-} is reduced to S_2^{2-} in the anode side; and the reverse occurs during discharge [28]. The open-circuit cell voltage is around 1.5 V, which depends on the active species, and operating cell voltage is around 1.36 V. The electrolyte solutions are separated by a cation-selective membrane, and the charge balance is maintained by transporting Na^+ through the

Table 2.1 Technical Comparison of all-vanadium VRB with other chemistries. Reprinted with permission from [28]. Theoretical specific energy numbers are shown in parentheses. Copyright (2011) American Chemical Society.

Type	OCV (V)	Specific energy (Wh/kg)	Discharge time (h)	Self-Discharge % per month at 20°C	Cycle life	Round-trip DC energy efficiency
VRB	1.4	15 (29)	4–12	5–10	5,000	70–80%
PSB	1.5	20 (41)	4–12	5–10	2,000	60–70%
ICB	1.18	<10	4–12		2,000	70–80%
ZBB	1.8	65 (429)	2–5	12–15	2,000	65–75%

membrane. Electrodes in PSB often include high-surface-area carbon [43, 44] and Ni foam [44].

ICB The ICB was the earliest storage device that used two fully soluble redox couples that were pumped through a power conversion cell. Systematic work was carried out by NASA on this system from 1970 to 1980 [42]. The Fe^{3+}/Fe^{2+} and Cr^{3+}/Cr^{2+} redox couples were employed in the catholyte and anolyte, delivering a cell voltage of 1.18 V. The cell reactions involve Cr^{2+} oxidizing to Cr^{3+} at the anode side and Fe^{3+} reducing to Fe^{2+} at the cathode side during discharge. ICB systems can operate with either a cation or an anion exchange membrane and employ carbon-based electrode materials [28]. The Fe^{3+}/Fe^{2+} couple shows high reversibility and fast kinetics, while the Cr^{3+}/Cr^{2+} couple shows a relatively slow kinetics on the electrodes. Hence, catalysts (Au, Pb, Bi) were often required for the anolyte to enhance its kinetics. In addition, the catalysts also yield a high overpotential for hydrogen evolution to suppress the H_2 production, which is a competing process with Cr^{3+} reduction to Cr^{2+} during charging. The cross-transport of iron and chromium ions is suppressed using mixed electrolytes at both cathode and anode sides, which requires the use of a microporous separator and operating temperature at 65°C [45].

2.2.6 Current Status and Practical Challenges

VRB The VRB is the most mature flow battery technology and accounts for 75 MWh of deployed systems [46]. The current technologies are still expensive in capital cost and life-cycle cost. VRBs are the most expensive flow battery chemistry, forecasted to cost $516/kWh in 2024 based on a model developed by Lux Research. RFB developers claim that sourcing vanadium from fly ash could reduce costs from over $500/kWh today to $300/kWh at scale. However, it has been estimated that even in the unrealistic scenario of a free vanadium electrolyte, VRB system costs will be $324/kWh in 2024 [46]. The Energy Storage Technology and Cost Characterization Report from DOE projected an even higher cost at $425/kWh in 2025 [47]. Recent research has suggested that improving the power density of VRB will drive down costs. Improvements in cell stack power density, for example, can cut costs by 33% [32, 46, 48].

ZBB Integrated ZBB energy storage systems have been tested on transportable trailers (up to 1 MW/3 MWh) for utility-scale applications. Multiple systems of this size could be connected in parallel for use in much larger applications. ZBB systems are also being supplied at the 5 kW/20 kWh Community Energy Storage (CES) scale, and are now being tested by utilities, mostly in Australia [49]. Practical challenges are associated with the high cell voltage and highly oxidative Br_2, which demand expensive cell electrodes, membranes, and flow-controlling components that have high chemical stability. Additionally, due to the high toxicity of Br_2 through inhalation and absorption, a stable amine complex is key to ZBB system safety, which requires an active cooling system to maintain stability. Furthermore, repeated plating of Zn in general can induce formation of dendrites, which might penetrate the separator, thus leading to cell failures. Controlled operating modes, such as pulsed discharge during charge, are often required to achieve uniform plating and reliable operation.

PSB The PSB technology has been one of the largest scale-up RFB systems owing to their moderate cost of electrodes and low cost of the electrolytes. A 1 MW pilot plant at Abershaw power state in South Wales was demonstrated and successfully tested. The construction of 15 MW/120 MWh demonstration plants started at Little Barford in the early 2000s [45]. A number of technical challenges still exist for PSBs: (i) there is a risk of electrolyte crossover through the membrane; (ii) long-term transport of species across the membrane requires analytical monitoring and chemical treatment of the electrolytes, increasing cost and complexity of the process plant; (iii) the system also encounters sulfur loss during extended cycling due to the buildup of sulfur species (e.g. S^{2-} and/or HS^-) on either the electrodes or membrane; (iv) mixing of the electrolytes can generate heat and toxic gases such as Br_2 and H_2S.

ICB From 1973 to 1982, a 1 kW/13 kWh ICB system was designed, fabricated, and tested by NASA. However, the slow kinetics and cross-contaminations of the electrolytes halted the system-level efforts. 10 kW and 60 kW system prototypes were manufactured and tested during 1984 to 1989 in Japan. The ICB was studied for energy storage in wireless telecom applications by Deeya Energy® in Silicon Valley, USA, but the company switched to VRB later due to the hydrogen bubbles formed in the electrolyte of ICB, restricting flow and reducing the current [50].

Nonaqueous RFBs While still in early development, nonaqueous RFBs are rapidly growing technologies. The major challenges associated with aqueous inorganic RFBs are (i) the limited electrochemical window to a maximum of 2.1 V, in which the redox pairs can function largely without water decomposition forming H_2 and O_2, (ii) side reactions of the redox pairs with the water, and (iii) high costs of the active materials. The specific limitation of the voltage window led to investigation on the use of nonaqueous alternative solvents and redox pairs. To date, research work including vanadium-based organic acetylacetonate complex in acetonitrile [51], nonaqueous chromium [52], manganese-based [53] and lithium-based RFBs [54], organic RFBs with organic redox pairs and organic solvent [55] or water [56] have been described, demonstrating wide voltage windows, opportunity for multi-electron transfer, and high energy density. However, the transfer of the results from research to commercialization is not easy, since often only small concentrations and quantities of active materials are used in research-based batteries. In real commercial batteries, organics flammability, propensity for side reactions, and deposition on electrodes remain as challenges. Although the advantages are clear, significant research and development activity are needed to enable a future practicable use of organic RFBs.

2.3 High-Temperature (HT) Sodium Batteries

2.3.1 Sodium–Sulfur (NaS) Batteries

Lithium–sulfur (Li–S) batteries are considered as promising large-scale energy storage systems due to their high specific capacity and energy density (1,675 mAh/g and 2,600 Wh/kg) [57]. However, taking into account sustainability and economic viewpoints, sodium is a more intriguing candidate owing to its natural abundance, which inspired research efforts for the Na–S battery [58]. The first Na–S battery was proposed in the 1960s by Ford Motor followed by NASA, and required a high operating temperature (>300°C) with molten electrodes and a solid beta-alumina electrolyte [59]. The high temperature is required because the polysulfide melt solidifies below 280°C and the ionic conductivity of the solid electrolyte is heavily compromised at lower temperatures [60]. A commercialized tubular HT Na–S system was achieved by Tokyo Electric Power Company (TEPCO) and NGK Insulator Ltd in 2002.

2.3.2 Reaction Mechanism

For the Na–S system, the sulfur cathode along with the sodium anode can deliver a theoretical energy density of 760 Wh/kg, which is two times higher than that of Pb–acid [60]. In terms of capacity, a complete discharge of elemental sulfur to sodium sulfide corresponds to a conversion reaction with two electrons per sulfur atom with a theoretical capacity of 1,672 mAh/g. However, the reversibility of the system is in peril due to the insoluble nature of the lower polysulfides (Na_2S_x, $x < 3$) at the battery's operating temperature, and therefore a more practical capacity of 558 mAh/g at the sodium trisulfide (Na_2S_3) state of formation during discharge is often reported [61].

The major components of the HT Na–S batteries are the beta-alumina solid electrolyte (BASE), the electrodes of sodium and sulfur in liquid state, and the container. During discharge, Na ions migrate through the BASE to the sulfur cathode to form sodium polysulfide intermediates and electrons flow in the external circuit (Figure 2.5a). Upon subsequent charge, Na ions will be released from the sodium polysulfides and recombine as elemental sodium at the anode. Negligible ohmic polarization is induced by contact resistance at the Na| β"-Al_2O_3 interface owing to the intimate contact between the molten Na and BASE at high temperature. The cell has an electromotive force of 2.076 V in the two-phase region when the starting anodic and cathodic reactants are in their pure elemental forms (Figure 2.5b). The operational cell potential window is between ~2.1 and 1.8 V, which is dependent on temperature and melt compositions. At the cell operating temperature, the melt is completely ionized, containing sodium and sodium polysulfides. In Na_2S_x, at the sulfur-rich state (x > 5), two liquid phases are present in the melt, which can solidify at 253°C for the Na_2S_5 and at 115°C for β-S_8, respectively. In general, the capacity loss stems from the insufficient charging at the two-phase zone. During the early stage of discharge (Eq. (2.10)),

$$2Na + 5S \rightarrow Na_2S_5 \quad E = 2.076 \text{ V at } 300°C, \quad \text{(Eq.(2.10))}$$

the reaction begins adjacent to the Na| β"-Al_2O_3, and the two-phase zone gradually propagates through the catholyte to the current collector side until complete filling of the cathode with Na_2S_5. Deeper discharging will result in the formation of Na_2S_4, as illustrated in Eq. (2.11):

$$2Na + 4Na_2S_5 \rightarrow 5Na_2S_4 \quad E = 1.970 \text{ V at } 300°C, \quad \text{(Eq.(2.11))}$$

The melting point of Na_2S_4 lies at 290°C, whereas polysulfide melts with compositions of Na_2S_x (3 < x < 5) solidify at lower temperatures (< 290°C). The reaction proceeds until Na_2S_3 is achieved, according to Eq. (2.12), which leads to the decrease of EMF to 1.78 V:

$$2Na + 3Na_2S_4 \rightarrow 4Na_2S_3 \quad E = 1.74 - 1.81 \text{ V at } 300°C. \quad \text{(Eq.(2.12))}$$

Continued discharge to Na_2S_2 is prohibited due to the negative Gibbs free energy of formation based on the density functional theory (DFT) calculations [62]. In practical HT Na–S cells, it is postulated that the governing carrying species are the sodium ions [60].

2.3.3 Cell Components

Beta-Alumina Solid Electrolyte (BASE) The key requirements for a functional solid electrolyte of Na–S batteries include a high Na-ion conductivity with minimal electron conductivity, high chemical stability toward sulfur species, impermeability, and sufficient thermal and mechanical strength. Beta-alumina species are the

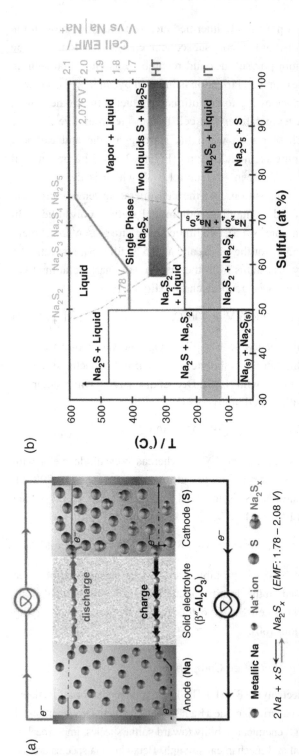

Figure 2.5 (a) Diagram depicting the operation of the Na–S battery. (b) Phase diagram of the Na–S system for the commercialized HT system. Temperature versus atomic percentage of sulfur. Figures are adapted from ref [60]. Published by The Royal Society of Chemistry.

building blocks of the BASE, which can be synthesized via several techniques including sol–gel, solid-state reaction, coprecipitation, and spray-freeze/freeze-drying [28]. Traditional solid-state synthesis of beta-alumina includes reaction of α-Al_2O_3 with Na_2O in the presence of Li_2O and/or MgO as stabilizers, which will form the product with a general formula of $Na_{1+x}Mg_xAl_{11-x}O_{17}$ ($x \approx 0.67$). For sodium beta-alumina with a general structure of $Na_2O \bullet xAl_2O_3$, depending on x, two distinct crystal structures are formed, beta-alumina (β-Al_2O_3, x = 8–11) and beta''-alumina (β''-Al_2O_3, x = 5–7). The major difference between these two materials stems from the chemical stoichiometry, sodium ion conductivity as well as stacking sequence of O^{2-} across the conduction layer. Specifically, the β''-Al_2O_3 possesses a rhombohedral structure, which is 50% larger than that of β-Al_2O_3, thus leading to enhanced ionic conductivity and representing a more suitable BASE in the case of the HT NaS cell. The diffusion of ions through the β''-Al_2O_3 is a thermally activated process. The surface of the electrolyte in the cell is partially wetted at 300°C, and it becomes fully wetted at 350°C. The effective sodium ion transport in BASE renders HT NaS a competitive energy storage system. However, it is largely dependent on temperature and is prone to form blocking layers upon long-term cycling and exposure to air. In addition to BASE, sodium (Na) super ionic conductor (NASICON) is widely used in Na-ion batteries. Recently, NASICON type Na-ion solid-electrolyte membrane, $Na_3Zr_2Si_2PO_{12}$, was investigated in room-temperature Na–S batteries [63].

Sulfur Cathode Elemental sulfur at 300–350°C is a viscous liquid that is reduced to polysulfide ions upon discharge and re-forms on charge. Initial discharge forms Na_2S_5, which has a higher density (1.86 g/cm³) than sulfur (1.66 g/cm³). Subsequent discharge generates Na_2S_3 (1.87 g/cm³), which has a similar density as Na_2S_5. The conductivity of sulfur at 300°C is 10^{-8} S cm^{-1}, which is much higher than that of 5×10^{-30} S cm^{-1} at 25 °C. Diffusion coefficients of Na_2S_5 at 350°C linger between 1×10^{-5} and 6×10^{-7} cm²/s [64]. High electrical currents are achievable under the provision of high electrode surface area and solubility of sulfur in the polysulfide at the electrolyte/melt interface. Owing to the highly corrosive environment, current collectors and cell containers are limited to molybdenum, chromium, aluminum, and stainless steel. Carbon materials have also been commonly explored as hosts for sulfur due to their high conductivity, chemical stability, and large surface area. For example, graphite felts and cloths with oriented fibers have been used as common cathodic collectors [65]. To mitigate for the poor wettability of sulfur on carbon felt, layered structures containing carbon fibers accompanied by a resistive layer are placed perpendicularly to the BASE, creating pathways for the sodium ions to reach the anode while acting as a sulfur deposition suppressant at the same time [66]. In order to enhance polysulfide solubility and inhibit undesirable discharge products such as sulfur deposition on the BASE, additives such as selenium, SiO_2, and metal sulfides have been explored [60]. In general, the critical requirements for the sulfur electrode include good corrosion resistance, high conductivity, good wettability against the polysulfides formed during the cell operation, and low weight and cost.

Sodium Anode The high reactivity of the molten sodium can afford high current densities ranging from 40 to 300 mA/cm^2, rendering it a critical component for the robust operation of the HT Na–S cell [67]. When Na is consumed during discharging, a volume decrease occurs. It is of critical importance to maintain close contact between Na and the BASE, thereby minimizing anode resistance. Several strategies have been adopted to achieve this goal including (i) feeding sodium by gravity from a top reservoir, (ii) wicking sodium to the BASE surface, and (iii) forcing sodium from a reservoir by gas pressure [68]. Additionally, interfacial polarization at the Na/BASE interface can be suppressed by constantly replacing the sodium electrode, treatment of the ceramic surface to improve wetting, and introduction of oxygen scavengers such as titanium and aluminum into liquid sodium. In order to devise a sodium electrode, the electrolyte separator must be thinner than the equivalent beta-alumina membrane to achieve the low resistance.

2.3.4 Cell Design

Tubular and planar cells are the prevailing designs for Na–S batteries, which exhibit different bulk transport mechanisms of their active species to the reaction front formed on the outside of the BASE. Planar cells (Figure 2.6a **and c**) offer several distinct advantages including efficient stacking, direct intercell connection, better stability, and larger active area of BASE per unit weight of the cell. The planar cells are widely used for fundamental battery chemistry studies such as wetting characteristics of molten sodium on the BASE, in situ monitoring of polysulfides, as well as testing of novel cathodes. Furthermore, non-glass sealings can also be implemented here due to the fewer joints and simpler cell architecture. A previous study [65] suggested that the deviation from the ideal open-circuit voltage, especially during discharge, was much smaller in the planar cell when compared to the tubular cell, which renders more stable and high-power operation by reducing power loss. Despite the numerous advantages, no practical operations have been achieved with planar cells due to thermomechanical fracture in the cell joint and/or BASE region. Additionally, volume changes during charge and discharge on either side of the BASE membrane can pose problems for this design as well.

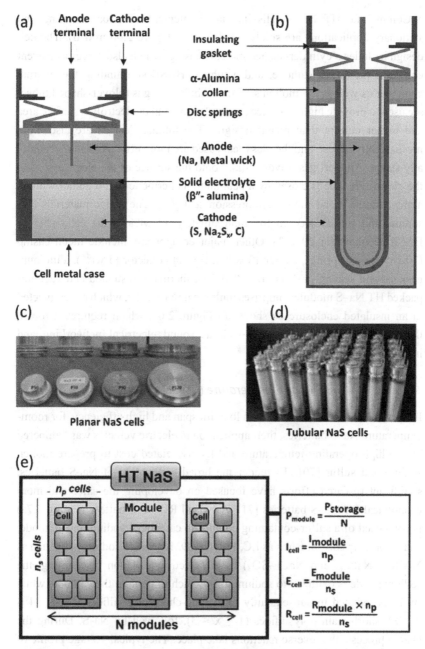

Figure 2.6 Schematic components of (a) planar and (b) tubular Na–S cells. Optical pictures of planar (c) and tubular (d) Na–S cells. (e) Na–S assembly of modules and cells with the respective equations describing its parameters (E: potential, P: power, I: current, R: resistance). Figure is reproduced from ref [60]. Published by The Royal Society of Chemistry.

Commercial HT Na–S cells that are implemented in load leveling and stationary applications are solely tubular cells (Figure 2.6b **and d**). The cell design includes a cylinder-shaped metallic casing, which also serves as current collector, the sulfur cathode, and a tubular BASE surrounding the internal container as well as the molten sodium. Tubular designs (clover-shaped tubes) are used to provide higher surface area and allow higher electrolyte surface area and higher current than planar designs. The tubular designs are also more mechanically robust, and the glass seals used are more effective and mechanically stable. An inert gas drives sodium onto the surface of the electrolyte, and the electrolyte tube is closed by a beta-alumina hermetic seal together with an alpha-alumina metal thermocompression seal [69]. The active materials (i.e. sodium and sulfur) account for approximately 32 wt% of the cell, whereas BASE accounts for 20 wt%. Other major components include metal casing (30 wt%), disc springs, carbon (3 wt%), current collector (3 wt%), aluminum gaskets, and sealing gaskets (12 wt%). The thermally insulated and vacuum-packed HT Na–S module comprises multiple unitary cells, which are connected in an insulated enclosure as shown in Figure 2.6e, which requires a robust thermal management system as well as sound electrical networking and insulation.

2.3.5 Room Temperature (RT) Na–S Batteries

Despite the potential advantages in long lifespan and high efficiency, for room-temperature Na–S batteries, their application in electric vehicles was hampered by the high operating temperature and the associated cost to prepare molten sodium and sulfur [70]. To render the broad application of Na–S batteries, significant research efforts have focused on developing the safer and more economical RT Na–S batteries [71]. A typical RT Na–S battery (Figure 2.7a) is composed of a sulfur-containing composite cathode, a sodium metal anode, and an organic solvent (such as EC, PC, TEGDME) with a sodium salt (such as $NaClO_4$, $NaPF_6$, and $NaCF_3SO_3$) as the electrolyte. Upon discharging, the sodium anode is oxidized to sodium ions, which migrate to the cathode, while sulfur is reduced to subsequently form long-chain polysulfides (Na_2S_x, $4 \leq x \leq 8$), short-chain polysulfides ($1 \leq x \leq 3$), followed by Na_2S. During the charge process, the reverse reactions take place. The typical voltage profile of the RT Na–S battery is depicted in Figure 2.7b, which shows a first plateau at ~2.2 V (region I), corresponding to a solid–liquid transition from sulfur to Na_2S_8; the sloping region from 2.2–1.65 V (region II) represents liquid–liquid transition from Na_2S_8 to long-chain polysulfides. The second plateau at ~1.65 V (region III) can be attributed to a liquid–solid transition from Na_2S_4 to insoluble

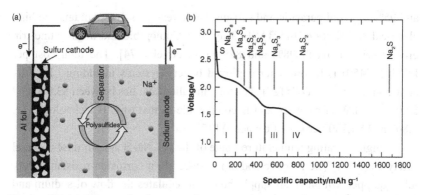

Figure 2.7 (a) Schematic diagram of RT Na–S batteries. (b) Theoretical versus practical discharge capacities of the Na–S cells operated at room temperature.
Reproduced with permission from ref [70]. Published by The Royal Society of Chemistry.

Na_2S_3 and Na_2S_2; the second sloping region at 1.65–1.2 V (region IV) represents the solid–solid reaction between Na_2S_2 and Na_2S. Previous reports have suggested that region II plays an important role in determining the cycling stability of the Na–S cell, as the long-chain polysulfides could dissolve into the electrolyte, leading to shuttle effects and capacity fade [70].

The mechanism of RT Na–S batteries is different from that at higher temperatures mainly due to the solid-state nature of sodium which leads to dendritic growth and to short circuit during charging [60]. In addition, sodium will form a nonuniform solid electrolyte interphase (SEI) layer upon contact with the electrolyte. RT Na–S technology does show promise as a reliable high-energy-density power source; however, intensive research efforts are required in addressing the compromised stability due to sluggish electroactivity, limited cycle life, rapid polysulfide migration, and self-discharge, in order to achieve large deployment.

2.3.6 Commercial Status and Safety

Today, Na–S battery technology is commercially available for grid applications, with some 200 installations worldwide, accounting for 584 MW of discharge power capacity, taking into account the US DOE Energy Storage Database and the leader of the NGK Insulators [27, 72], which is suitable for up to 8 hours of daily peak shaving and a storage capacity of 3,700 MW h. A stable capacity of HT Na–S batteries can be achieved for 4,500 cycles to 80% depth of discharge at a discharge rate of 8–10 hrs, which corresponds to a calendar life of at least 15 years [73]. Generally, the state of charge of the battery is controlled between 20 and 95% to avoid parasitic and self-discharge reactions and maintain the high electrical resistivity of the BASE. Typically, Coulombic and mean round-trip efficiencies can reach 98% and 82%, respectively [74]. The total cost per kW h (~$450) is below other deployed battery systems including LIB, Pb–acid, and redox flow cells [27]. The total installed cost lies between $3,000 and $4,000 per kW of capacity and is comparable to that of LIB technology (i.e. 1,800 and $4,100 per kW of capacity) [75].

The high operating temperature range of the HT Na–S system raises several safety concerns. The reactive sodium is stored in a corrosion-resistant safety tube equipped with a small supply hole that regulates the flow of sodium and also allows a minimum amount of sodium to maintain the electrochemical reaction. Sulfur is not chemically active but is highly flammable and generates toxic sulfur dioxide gas. Hence, the operating temperature of HT Na–S should not exceed 400°C under any circumstance. In addition, due to the hygroscopic nature of polysulfides, a water-free environment is a prerequisite, which requires inner and outer protection sheets to be added for thermal insulation and fire resistance. Current cell designs utilize a chromium-containing layer as the primary corrosion barrier followed by stainless steel, to mitigate the radial stress concentration during the fluctuating operating temperature of this energy storage system. Furthermore, cracking of the BASE due to nonuniform current distribution leading to high local current density degradation of the glass seal, corrosion of the container, and breaching of the metal-to-ceramic seals could cause cell failure. Therefore, rigorous safety testing protocols are required involving self-extinguishing modules with careful consideration of container design, container drop, fire resistance, flood, module drop, and the mitigation of propensity toward external short circuits [60].

2.3.7 Sodium Metal Halide (ZEBRA) Batteries

ZEBRA (Zeolite Battery Research Africa Project) batteries are another form of high-temperature sodium battery that utilizes a molten sodium anode separated from the cathode by BASE [76, 77]. The original design was proposed by Coetzer [78], and today is primarily produced by FZ SoNick out of Switzerland [76]. Typically, ZEBRA batteries contain a $NiCl_2$ cathode, which has advantages over the sulfur cathode with a higher operating voltage of 2.58 V versus Na/Na^+ and improved safety. Due to the extensive similarities between the Na–S system and the ZEBRA system, this section will focus on distinguishing attributes of ZEBRA batteries.

Reaction Mechanism In the standard $Na-NiCl_2$ battery, the cathode consists of solid Ni metal particles and sodium chloride suspended in a molten

$NaAlCl_4$ catholyte [76, 77]. The melting point of $NaAlCl_4$ is 157°C, which is the lowest possible operating temperature for the ZEBRA batteries; however, as discussed in Section 2.3.3, to decrease the resistance of BASE, an operating temperature of 300–350°C is typically used. Upon discharge, the reaction proceeds as follows:

$$2Na + NiCl_2 \rightarrow 2NaCl + Ni \quad E = 2.58 \text{ V at } 300°C, \quad\quad (Eq.(2.13))$$

This reaction results in a marginally higher theoretical energy density of 788 Wh/Kg than the NaS battery, but a lower theoretical capacity at 305 mAh/g [60, 76]. Although the theoretical capacity is lower, the observable capacities of the two HT sodium systems are similar.

A unique feature of the $NaAlCl_4$ catholyte is that it can be reversibly used to overcome issues with overcharging. This occurs via the following reaction [76, 77]:

$$2NaAlCl_4 + Ni \rightarrow 2Na + 2AlCl_3 + NiCl_2 \quad E = 3.05 \text{ V at } 300°C,$$
$$(Eq.(2.14))$$

The ability to reversibly overcharge relates to an excess of Ni inhibiting gaseous chlorine formation. This is particularly useful for multicell battery packs used in grid energy storage, because if one cell in a parallel string fails, the other cells can be overcharged to compensate. This feature is unique to the ZEBRA system and is not possible for the Na–S batteries.

Metal Chloride Cathode At 300°C in the discharged state, the typical ZEBRA cathode contains solid particles of Ni and NaCl in a liquid $NaAlCl_4$ catholyte and, upon charging, $NiCl_2$ forms on the Ni particles as Na^+ transfers through the BASE. The main cause of capacity fade, however, comes from the particle sizes of the Ni and NaCl [76, 77]. With larger particles, incomplete conversion occurs upon charging due to passivation of Ni particles with a $NiCl_2$ surface layer. To control the particulate sizes, a small amount of sulfur in the form of FeS is typically added, which poisons the Ni surface, making it difficult for excessive Ni buildup on the Ni particles during discharge. Other sodium halides are added to the mix to help remove Ni and Fe buildup on the surface of BASE, thereby preventing an increase in cell resistance over time [76, 77]. Recently, a porous conductive carbon matrix has been used as a cathode current collector [79, 80]. In these reports, the improved conductivity and physical constraints aided in inhibiting particle growth, thereby prolonging the battery life without the need for additives.

ZEBRA batteries are not exclusively Na-NiCl$_2$, as other metal chlorides can also undergo the same reactions. The most common alternative to Ni is Fe [81, 82]. The Na-FeCl$_2$ battery undergoes a similar reaction to Eq. (2.14) and upon discharge proceeds as follows:

$$2Na + FeCl_2 \rightarrow 2NaCl + Fe \quad E = 2.35 \text{ V at } 300°C, \qquad (\text{Eq.}(2.15))$$

The reaction with Fe does possess a lower voltage; however, the severe decrease in cost and improved kinetics are desirable [82]. Na-ZnCl$_2$ has also been suggested as an inexpensive alternative to Na-NiCl$_2$ batteries; however, the voltage is even lower than Fe (1.94 V, 2.13 V vs. Na/Na$^+$) [83, 84].

Cell Design Similar to the Na–S cell, a tubular format is used for commercial ZEBRA batteries. In the ZEBRA cells the placement of the cathode and anode compartments is switched; however, a schematic is shown in Figure 2.8.

In the cathode compartment, a highly porous Ni current collector is used to aid the longitudinal conduction to the cell terminal [77]. Current manufacturing utilizes a clover-shaped BASE instead of a circular-shaped BASE, which increases surface area and decreases resistance. To help improve the wetting of the molten Na anode to BASE, various strategies have been utilized including thin surface layers of Pb, Bi, or Sn on BASE; Ti and Al oxygen scavengers in the Na melt; Na-K, Na-Rb, and Ni-Cs molten alloy anodes; and addition of wicking metal grids to the BASE exterior [85]. During manufacture, one of the best features of the ZEBRA-type batteries is that the cells can be assembled in the fully discharged state using metal powders and NaCl [77]. This avoids working with elemental Na during processing, and upon completing construction, fully charging the cell will convert the components into the desired air-sensitive NiCl$_2$ and Na metal.

Commercial Status Na-NiCl$_2$ battery technology is commercially available from FZ SoNick and has a similar extent of deployment as the Na–S battery. Unlike the 8 hours of discharge that Na–S batteries provide, the Na-NiCl$_2$ batteries constitute only ~3 hours of discharge at 80% depth of discharge [86]. Na-NiCl$_2$ batteries have a calendar life of at least 15 years and can hold a stable capacity for 4,500 cycles. The total cost per kWh in 2015 was ~$900, which is double that of the Na–S battery [86]. Although the costs are currently high for this technology, large cost reductions could be achieved by replacing NiCl$_2$ with FeCl$_2$. Additionally, the limiting temperature is 157°C of the molten NaAlCl$_4$, which is much lower than the required polysulfide temperatures of 280°C; therefore, intermediate-temperature ZEBRA batteries are possible. Through the use of planar cells with thin BASE discs, many current literature

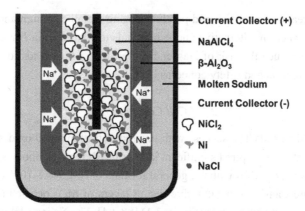

Current Collector (+)

NaAlCl$_4$

β-Al$_2$O$_3$

Molten Sodium

Current Collector (-)

NiCl$_2$

Ni

NaCl

Figure 2.8 Schematic of a tubular ZEBRA Na-NiCl$_2$ cell.

reports indicate the feasibility of operation at 190°C, which would constitute additional cost reductions in cell design and continual heating costs [79, 80, 82, 83, 85].

3 Potential Battery Technologies for Grid-Scale Electrical Energy Storage

3.1 Na-Ion Batteries

3.1.1 Overview

With the success and effective widespread implementation of Li-ion batteries, it is natural to look toward Na-ion batteries as a potential replacement. Both Li and Na belong to the alkali metals group and have very negative reduction potentials (−3.04 V vs. −2.71 V vs. SHE), similar (de)insertion chemistries, and due to their similar physical properties, the infrastructure in place for Li-ion battery processing and manufacturing can likely be utilized for Na-ion batteries. The big advantage of Na-ion batteries is the vast abundance of non–geographically constrained sodium, leading to significant cost reduction compared to lithium [1, 87]. The increased mass of Na-ion batteries leads to a lower energy density than Li-ion batteries. Therefore, Na-ion is unlikely to replace Li-ion for portable electronic applications; however, for stationary grid energy storage application, Na-ion battery could be a very attractive alternative. There remain several challenges to overcome before Na-ion technology will be ready for large-scale commercialization. The larger ionic radius of Na$^+$ (+25–55%), relative to Li-ion depending on the coordination) impedes intercalation into typical Li-ion cathode materials, causing large volume expansions and sluggish kinetics [88, 89]. The lower melting point of sodium (98°C) relative to lithium

(181°C) creates greater risk and precludes using metallic sodium anodes [89]. Further advances in alternative anode materials to Na metal for Na-ion will be required to reduce the risk of Na dendritic growth, which leads to decreased capacity retention and safety concerns [87].

3.1.2 Cell Components

Cathode Materials The success with layered transition metal oxides (TMOs) in Li-ion batteries inspired significant study of TMOs as a cathode system for Na-ion batteries. TMOs with the general formula of $Na_{1-x}MO_2$ ($0 < x \leq 1$, M is one or more transition metals) give rise to layers of MO_6 octahedra between which NaO_x polyhedra reside (Figure 3.1) [88, 94]. The Na ion (de)intercalates between the MO_6 layers and gives rise to high capacities and voltages. The larger ionic radius of Na^+ compared to Li^+ (1.02 Å vs. 0.76 Å), results in larger volume expansions, which leads to poor long-term cycling stability [88, 94, 95]. By careful transition metal selection and combination of multiple transition metals, improvements have been made to capacity (150 – 200 mAh/g), voltage (3 – 4 V vs. Na/Na$^+$), and cycle stability (>70% over several hundred cycles). Several examples of TMOs used in Na-ion batteries are presented in Table 3.1. Although TMOs offer promising voltages and reasonable capacities for Na-ion batteries, significant strides in capacity retention and rate capability are still required. In addition to layered TMOs, MnO_2 possesses multiple tunnel structures, which can accommodate Na$^+$ intercalation; these tunnel structures will be discussed in more detail in the rechargeable magnesium battery and rechargeable aqueous zinc battery sections. The MnO_2 structures benefit from high capacities (~200 mAh/g), but lower voltages (1 – 2.5 V vs. Na/Na$^+$) and capacity retention (50–70% after 100 cycles) [96, 97]. The addition of Ag into the tunnels can increase Na insertion, improving capacity, but at the cost of greater capacity fade [98, 99].

To avoid the large volume expansion associated with intercalation into the layered structures of TMOs, framework-type cathodes have been investigated. Polyanionic-type cathodes consist of covalent open frameworks of anionic polyhedra (Figure 3.1). Many different structures and anions fall into this class of compounds such as orthophosphates $Na_3V_2(PO_4)_3$, pyrophosphates $NaVP_2O_7$, fluorophosphates $NaVPO_4$ F, sulfates $Na_2Fe_2(SO_4)_3$, and more [49, 87, 94]. The covalent polyhedral networks form channels with low-energy Na$^+$ migration pathways, small volume changes, and good thermal and oxidative stability [49]. Additionally, the inductive effects from the polyanions allow for some of the highest operating voltages (3.5 – 4 V vs. Na/Na$^+$) of any Na-ion cathode material, which helps the polyanion cathodes to obtain a high specific

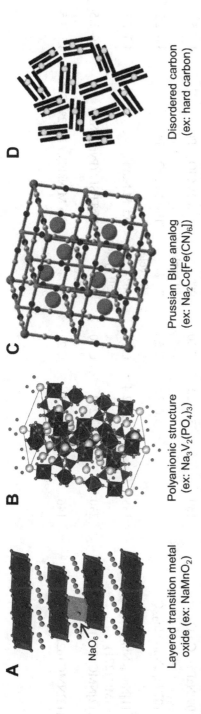

A

Layered transition metal oxide (ex: NaMnO$_2$)

NaO$_6$

B

Polyanionic structure (ex: Na$_3$V$_2$(PO$_4$)$_3$)

C

Prussian Blue analog (ex: Na$_2$Co[Fe(CN)$_6$])

D

Disordered carbon (ex: hard carbon)

Figure 3.1 General crystal structures of (a) layered transition metal oxide cathodes reprinted with permission from ref [90], published by The Royal Society of Chemistry; (b) polyanionic cathodes [91]; (c) Prussian Blue analog cathodes reprinted with permission from ref [92], copyright (2016) American Chemical Society, and a schematic of Na$^+$ intercalated into hard carbon reprinted with permissions from ref [93], copyright Elsevier (2018).

Table 3.1 A collection of various reports on cathode materials for Na-ion batteries. Included are capacity at a low and high rate (where applicable), approximate voltage plateaus (V vs. Na/Na$^+$), charging rate associated with the two capacities, capacity retention over the stated number of cycles, electrolyte used, and the voltage range over which the battery was cycled at. All reports used a Na metal anode in the battery configuration.

Material	Capacity (mAh/g)	Voltage	Rate	Retention	Electrolyte	Voltage Range
[107]$Na_{0.7}CoO_2$	125	3.6–2.3	0.4 C	86% 300 cycles	1 M $NaClO_4$ / 1:1 EC:DEC	3.8–2
[108]$NaCrO_2$	120	3.3, 2.9	25 mA/g	75% 50 cycles	1 M $NaClO_4$ / PC	3.6–2
[109]$NaFe_{1-y}Co_yO_2$	160	3.5, 2.8	C/20	90% 50 cycles	1 M $NaClO_4$ / PC	4–2.5
[110]$Na_{1-x}Ni_{0.5}Mn_{0.5}O_2$	125	3.6–2.5	C/5	75% 50 cycles	1 M $NaClO_4$ / PC	3.8–2.2
[111]$Na_{0.45}Ni_{0.22}Mn_{0.66}Co_{0.11}O_2$ NMC (271)	135	4.2–2.4	C/10	70% 250 cycles	0.5 M $NaPF_6$ / PC	4.3–2.1
[112]NMC (271)	230	4.2–2	C/10	80% 100 cycles	NaTFSI / ($Pyr_{14}FSI$)	4.6–1.5
[113]NMC (622)	146	4–2.4	C/2	80% 50 cycles	0.5 M $NaPF_6$ / PC 2% FEC	4.1–1.5

Material	Capacity (mAh/g)	Voltage range (V)	Rate	Capacity retention	Electrolyte	Voltage window (V)
[114]$NaNi_{0.5}Ti_{0.5}O_2$	121, 72	4–2.9	C/5, 5 C	>90% 100 cycles	1 M $NaPF_6$ / 1:1 EC:DMC	4.7–2
[115]$NaNi_{0.45}Cu_{0.05}Mn_{0.4}Ti_{0.1}O_2$	124	3.7–2.6	1 C	72% 500 cycles	1 M $NaClO_4$ / PC 5% FEC	4–2
[116]$NaTi_{0.25}Fe_{0.25}Co_{0.25}Ni_{0.25}O_2$	163	3.8–2.5	C/20	89% 20 cycles	1 M $NaPF_6$ / 1:1 EC:DEC	4–2
[96]α-MnO_2	200	2.5–1	20 mA/g	38% 100 cycles	1 M $NaClO_4$/ 1:1 EC:PC	4.3–1
[97]β-MnO_2	300	2, 2–1	20 mA/g	64% 100 cycles	1 M $NaClO_4$/ 1:1 EC:PC	4.3–1
[117]β-Na_xMnO_2	185	3.5–2.5, 2.5	C/20	70% 100 cycles	1 M $NaPF6$ / 45:45:10 EC: PC:DMC	4.2–2
[98]$Ag_{1.22}Mn_8O_{16}$	247	2.7–2.3, 1.7–1.3	20 mA/g	11% 50 cycles	1 M $NaPF_6$ / 1:1 EC:DEC	3.8–1.3
[118]$Na_2Fe_2(SO_4)_3$	100, 50	4–3.5	C/20, 20 C	>90% 30 cycles	1 M $NaPF_6$ / 1:1 EC:DEC 2% FEC	4.5–2
[119]NASICON $Na_3V_2(PO_4)_3$	113, 91	3.4	1 C, 20 C	20 C: 89.7% 3,500 cycles	1 M $NaClO_4$ / 1:1 EC:PC	3.9–2.3
[120]$Na_{1.5}VPO_{4.8}F_{0.7}$	~158	4, 3.6	1 C	84% 500 cycles	1 M $NaClO_4$ / PC	4.7–2

Table 3.1 (cont.)

Material	Capacity (mAh/g)	Voltage	Rate	Retention	Electrolyte	Voltage Range
[121]$Na_3V_2(PO_4)_3$	180	4, 3.5, 1.5	C/20		1 M $NaPF_6$ / 1:1 EC:DEC	4.4–1.1
[122]$NaFePO_4$	140, 65	2.8	C/10, 10 C	10 C: 70% 6,000 cycles	1 M $NaClO_4$ / 1:1 EC:PC	4–2
[123]$Na_3V_2(PO_4)_3$	117, 90	3.4	C/5, 30 C	30 C: 75% 20,000 cycles	1 M $NaClO_4$ / 1:1 EC:DEC	3.9–2
[124]$NaTi_2(PO_4)_3$	120, 105	2.3–2.1	1 C, 10 C	10 C: 75% 10,000 cycles	1 M $NaClO_4$ / 1:1 EC:DEC 2% FEC	2.8–1.5
[125]$Na_2FeP_2O_7$	78	3.2–2.9	100 mA/g	91% 1,000 cycles	NaFSA / KFSA	4–2.5
[104]$FeFe(CN)_6$	120, 110	3.4, 2.8	C/2, 2 C	2 C: 87% 500 cycles	1 M $NaPF_6$ / 1:1 EC:DEC	4–2
[126]$Cu_3[Fe(CN)_6]_2$	44	3.5–3.2	20 mA/g	57% 50 cycles	1 M $NaPF_6$ / 1:1 EC:DMC	4.3–2
[127]$Na_{1.96}Mn[Mn(CN)_6]_{0.99}$ *$2H_2O$	209, 172	3.6, 2.8, 1.8	C/5, 2 C	2 C: 75% 100 cycles	1 M $NaClO_4$ / PC	4–1.3

[128]Na$_{1.60}$Co[Fe(CN)$_6$]$_{0.90}$*2.9H$_2$O	139, 121	3.8, 3.4	0.6 C, 60 C	0.6 C: 71% 100 cycles	1 M NaClO$_4$ / PC	4–2
[103]Na$_{1.89}$Mn[Fe(CN)$_6$]$_{0.97}$	150, 121	3.4	C/5, 20 C	0.1A/g: 75% 500 cycles	1 M NaClO$_4$ / 1:1 EC:DEC 10% FEC	4–2
[129]Na$_{0.84}$Ni[Fe(CN)$_6$]$_{0.71}$*6H$_2$O	66	3.3–3.1	20 mA/g	99.7% 200 cycles	1 M NaPF$_6$ / 1:1 PC/EC	4.1–2
[130]Na$_2$Zn$_3$[Fe(CN)$_6$]$_2$*xH$_2$O	56	3.6, 3.4–3.1	10 mA/g	85% 50 cycles	1 M NaPF$_6$ / 1:1 EC/DMC	4–2

energy [49]. The large covalent framework also allows for improved (dis) charging rates compared to TMOs, reaching reasonable capacities at rates over 20°C. The major drawback for these materials, however, is the high molecular weight of the polyanions, leading to low gravimetric capacities (80–150 mAh/g) [49]. The literature has addressed the conductivity and cycling rate by nanosizing the active materials and incorporation into a variety of conductive carbon structures. Several examples of polyanionic-type cathodes used in Na-ion batteries are presented in Table 3.1.

Another framework-type cathode is Prussian Blue, a metal organic framework, with the general formula $Fe^{3+}[Fe^{2+}(CN)_6]^-$ where the Fe^{2+} is octahedrally coordinated to the carbon of six CN ligands, which act as a bridge for the nitrogen to bond to Fe^{3+} (Figure 3.1) [100]. Replacement of one or both Fe ions results in Prussian Blue analogs (PBAs). PBAs are of great interest due to the open framework structure, which contains large ionic channels (3.2 Å diameter in the <100> direction) and interstitial sites (4.6 Å) [100]. These large channels and sites are ideal for the (de)intercalation and storage of the larger Na ion within the structure and has resulted in rapid ion diffusion ($10^{-8} - 10^{-9}$ cm²/s diffusion coefficients) and reports of long cycle lifetimes over 500 cycles at high capacity retention [101–106]. By tuning the transition metals in a PBA, a variety of capacities (50–200 mAh/g) and voltages (2.8–3.6 V) can be achieved. Several examples of PBA cathodes used in Na-ion batteries are presented in Table 3.1. Like the polyanionic-type cathodes, PBAs can cycle at higher rates than TMOs; however, the major drawback of PBAs is their structural stability. During the synthesis, water is easily coordinated by the structure leaking to clogged ionic channels, decreased electrical conductivity, and vacancy sites. This is believed to be the main cause of the poor capacity retention exhibited by PBA cathodes [100].

Electrolyte Conveniently, the extensive electrolyte studies for Li-ion systems seem to be highly transferable, and great strides in Na-ion electrolytes have been made with a series of high-conductivity, stable electrolytes already known. Some common examples of Na-ion electrolytes are $NaPF_6$ or $NaClO_4$ salts in combinations of propylene carbonate, ethylene carbonate, dimethyl carbonate, and diethyl carbonate. Because of the similarities between the Li-ion electrolytes and Na-ion electrolytes, Na-ion battery electrolytes will not be discussed in depth here. However, it is recognized that the alkali ion can impact solubility of the electrolyte salt, the (dis)charge products formed, and the stability of the electrode–electrolyte interfaces; therefore, additional studies toward understanding long-term stability of Na-ion systems will be needed to help the field progress toward commercialization and implementation.

Carbon Anode Materials In Li-ion batteries, graphite has become the dominant choice as an intercalation anode. For Na-ion batteries, however, the larger ionic radius of Na^+ precludes the insertion of sodium into graphite [88, 94, 131]. It was discovered that disordered carbon-type electrodes can accommodate the larger Na ion; in particular, hard carbon has received extensive investigation (Figure 3.1). The increased disorder of hard carbon improves electrochemistry for Na-ion insertion, resulting in capacities near 300 mAh/g [88, 131–133]. Initial studies with hard carbon exhibited poor capacity retention and large capacity loss on the first cycle [133]. Through electrolyte selection and nanosizing of the hard carbon particles, greater than 90% capacity retention (after the first cycle) over 50–100 cycles was observed [131, 132]. To further improve capacity and cycle life, other forms of carbon have been explored; see Table 3.2. Doped carbon nanofibers, doped porous carbon, graphene, and others have been implemented to increase capacity retention close to 100% with a cycle life in the hundreds to thousands of cycles [134–136]. Other types of anodes have been tested for Na-ion batteries including Na metal, alloys, and metal oxides. Many of the same issues that arise with these types of anodes in Li-ion batteries occur with Na as well, including Na metal dendritic growth, large volume expansion upon metal alloying, and metal oxide conversion leading to poor capacity retention [87, 88, 94]. As such carbon-based materials are likely to be the safest, most reliable, and cost-effective anodes for use in Na-ion batteries.

3.1.3 Current Status and Challenges

The type of Na-ion battery just described has not been currently implemented for any grid-scale energy storage applications. Aquion Energy created a different type of Na-ion battery than the one discussed here and deployed four instances of grid-scale aqueous Na-ion battery that utilized a MnO_2 cathode, a $NaTi_2(PO_4)_3$ anode, and 5 M $NaClO_4$ aqueous electrolyte [11, 13]. Since their installment, Aquion Energy has filed for bankruptcy. Austria's BlueSky Energy acquired the assets of Aquion and is now featuring a saltwater battery that can be deployed in residential installs. To progress nonaqueous Na-ion battery technology to the point of grid-scale commercialization, several challenges must be addressed. The selection of a cathode material that is capable of supporting higher capacities, rate capabilities, and cycle life are critical factors. Of the various types of cathodes discussed, none can meet all three requirements. Transition metal tuning in TMOs to improve capacity and rate, fluorination of polyanion cathodes to enhance capacity, and dehydration of PBAs to increase cycle life have all aided the progression of Na-ion batteries [103, 111–113, 120, 121]. With a suitable cathode material, and the

Table 3.2 A collection of various reports on anode materials for Na-ion batteries. Included is capacity at a low and a high rate (where applicable), approximate voltage plateaus (V vs. Na/Na$^+$), (dis)charging rate associated with the two capacities, capacity retention over the stated number of cycles, electrolyte used, and the voltage range the battery was cycled at. All reports used a Na metal counter electrode in the battery configuration.

Material	Capacity	Voltage	Rate	Retention	Electrolyte	Voltage Range
[131]Hard carbons	220	1.1–0.1	25 mA/g	100% 120 cycles	1 M NaClO$_4$, EC:DMC	2–0.01 V
[132]Hard carbons	290	1–0.1	20 mA/g	90% 50 cycles	1 M NaClO$_4$, EC:DMC	2–0.005 V
[136]Doped graphene	400, 260	1.4–0.2	0.03, 1 A/g	1 A/g: 100% 10,000 cycles	1 M NaClO$_4$/ 1:1 EC:PC 2% FEC	3–0.1 V
[135]N-doped carbon nanofibers	150	1.2–0.1	0.2 A/g	90% 200 cycles	1 M NaPF$_6$/ 4:4:2 EC:DEC:PC	2–0.1 V
[134]N/P-doped porous carbon	230	2.5–0.1	0.2 A/g	100% 700 cycles	1 M NaClO$_4$/ 1:1 EC:PC 5% FEC	3–0.1 V
[137]Expanded graphite	280, 180	1.7–0	0.02, 0.1 A/g	0.1 A/g: ~80% 2,000 cycles	1 M NaClO$_4$/ PC	2–0 V

Ref.	Material	Capacity (mAh/g)		Current density	Retention	Electrolyte	Voltage range
[138]	Carbon black	234, 120	1.25–0	0.05, 3.2A/ g	3.2A/g: 60% 2,000 cycles	1 M NaCF$_3$SO$_3$/ DEGDME	2.5–0.001 V
[139]	Amorphous carbon	284	1–0.1	0.1 A/g	91% 100 cycles	0.6 M NaPF$_6$/ 1:1 EC:DMC	2 – 0.01 V
[140]	Hard carbon	374	1.2–0.1	0.05 A/g	100% 200 cycles	1 M NaClO$_4$/ EC:DEC	3–0.01 V

already viable electrolytes and hard carbon anodes, Na-ion batteries could be commercially viable for grid-scale energy storage.

3.1.4 K-Ion Batteries

Similar to Na-ion batteries, K-ion batteries are a potential option for replacing Li, due to similar properties and lower cost. The K^+ ion has an even larger ionic radius of 1.38 Å, compared to Na^+ and Li^+ (1.02 Å and 0.76 Å) [95]. Although K^+ is larger, it maintains a lower Lewis acidity than Na^+ and Li^+, giving it the highest mobility in nonaqueous electrolytes [8, 141]. Additionally, it was determined that K^+ can intercalate into graphite, forming a KC_8 formula, which gives an improved and simplified anode over Na-ion batteries [142]. In spite of these benefits, the larger ionic radius of K^+ leads to more severe expansion issues upon cathode intercalation than Na^+ does, so many of the large open framework structures studied for Na-ion batteries have been tested for K-ion batteries. Although potassium is lower cost than lithium, it is still higher cost than sodium [141, 142]. These challenges must be addressed to make K-ion batteries a viable alternative for grid-scale electrochemical energy storage.

3.2 Rechargeable Magnesium Batteries (RMB)

3.2.1 Overview

In the pursuit of high-energy-density batteries for grid energy storage, consideration of divalent ion systems is highly attractive. For each ion utilized, two charges are gained, giving Mg metal a theoretical gravimetric capacity of 2,200 mAh/g and volumetric capacity of 3,835 mAh/cm^3 [94, 143–146]. In addition to high theoretical capacities, Mg is the fifth most abundant element in the crust, is more stable to ambient atmosphere than Li metal, typically forms less toxic compounds than Li, and has a reduction potential of −2.4 V versus SHE. Furthermore, unlike the alkali metals, dendrite-free platting and stripping can be achieved with Mg metal, allowing for its safe use as the anode in rechargeable Mg batteries [94, 146]. There are many challenges, however, for Mg batteries that must be over-come before Mg can be implemented on the industrial scale. Due to the 2+ charge and small size of the magnesium ion, strong electrostatic interactions occur, which cause desolvation issues in the electrolyte and problems with ionic transport and (de)insertion from the cathode [143–145]. Electrolyte decomposition on the Mg metal anode causes passivation layers, which drastically increases the overpotential for Mg plating [94, 144, 146].

Careful consideration of the electrolyte and cathode material are key for the future development of rechargeable Mg batteries.

3.2.2 Cell Components

Cathode Materials Due to the increased charge density associated with the 2+ charge of the Mg ion, magnesium diffusion through typical cathode materials used for Li-ion batteries is very sluggish and results in poor capacity retention. To accommodate the stronger electrostatic interactions, different types of materials have been explored, and several of the key materials that have been heavily investigated are discussed. Table 3.3 includes data for several selected examples of cathode materials and their performance for use in rechargeable magnesium batteries. Figure 3.2 shows crystal structures of three main classes of cathode materials discussed in this section.

The Chevrel phase intercalation cathode (Mo_6X_8, X = S, Se) has been the most thoroughly studied cathode material for rechargeable Mg batteries due to the large ion channels that provide fast intercalation kinetics for Mg^{2+}. The large ion channels have provided full depth of discharge with minimal capacity fade over thousands of cycles. Although the cycle life of the Chevrel phase cathodes is much better than all other materials, the low voltage plateau is only 1.1 V versus Mg/Mg^{2+}, and the theoretical capacity is only 122 mAh/g [152]. Additionally, because Mo_6S_8 or $Mg_2Mo_6S_8$ cannot be obtained directly, a long high-temperature synthesis of $Cu_2Mo_6S_8$ is required, followed by a multiday Cu leaching process [153, 154]. The combination of the low voltage, theoretical capacity, and expensive time-consuming production makes Chevrel phase intercalation cathodes impractical for grid-scale energy storage.

Another frequently investigated system is layered vanadium oxide, V_2O_5. V_2O_5 has a much higher voltage of >2.5 V and a theoretical capacity of 465 mAh/g with 2 equivalence of intercalated Mg^{2+}, making V_2O_5 a much more promising cathode material. Unfortunately, the Mg^{2+} intercalation is slow and the electrical conductivity is low, which have been compensated for by decreasing crystallite size and incorporation of conductive additives [143, 155–158]. Formation of V_2O_5 xerogels through hydration of the V_2O_5 structure causes the monolayer structure to become a bilayer structure with 0 to 3 moles of water per mol of V_2O_5 into the structure [159–161]. The water molecules can solvate the Mg^{2+} ion, partially shielding the high electrostatic charge, which significantly improves (de)insertion of Mg^{2+}. A comparison of pure V_2O_5 and V_2O_5 xerogel was performed and an increase from 50 mAh/g to 170 mAh/g was observed for the first cycle; however, subsequent cycling shows rapid capacity fade associated with loss of Mg^{2+} solvated water, resulting in structural collapse

Table 3.3 A collection of various reports on cathode materials for rechargeable magnesium batteries. Included is capacity at a low and a high rate (where applicable), approximate voltage plateaus (V vs. Mg/Mg^{2+}), (dis)charging rate associated with the two capacities, capacity retention over the stated number of cycles, electrolyte used, the anode used in the battery configuration, and the voltage range at which the battery was cycled.

Material	Capacity (mAh/g)	Voltage	Rate	Retention	Electrolyte	Anode	Voltage Range
[152]$Mg_2Mo_6S_8$	75	1.2, 1.1	0.3 mA/ cm^2	85% 2,000 cycles	0.25 M 1:2 Bu_2 Mg – $AlCl_2Et$ / THF	Mg metal	2–0.25
[175]$Mg_xMo_6S_8$	88	1.1, 1	C/8	100% 94 cycles	0.4 M 2:1 PhMgCl – $AlCl_3$ / THF	Mg metal	1.8–0.5
[176]$Mg_xMo_6S_6Se_2$	109	1.3, 1.1	C/8	>95% 100 cycles	0.25 M 1:2 Bu_2 Mg – $AlCl_2Et$ / THF	Mg metal	2–0.2
[156]V_2O_5/C	150	$-0.6 - -0.7$ vs. Ag/Ag^+	1 A/g		1 M $Mg(ClO_4)_2$ / AN	Mg metal	$0.2 - -1$ vs. Ag/Ag^+
[158]V_2O_5	180	2.25	0.5 µA/ cm^2	83% 35 cycles	0.5 M $Mg(ClO_4)_2$ / AN	Mg metal	3–2.2
[162]$V_2O_5 \cdot nH_2O$	170	~1 vs. Al/Al^{3+}		35% 20 cycles	$MgCl_2$ – $AlCl_3$ / EMIC	Al metal	1.8–0.3 vs. Al/Al^{3+}

Material	Capacity	Voltage	Current	Capacity retention / cycles	Electrolyte	Counter electrode	Voltage range
[163]$Mg_{0.1}V_2O_5$	300	$\sim$0--0.7 vs. Ag/Ag$^+$	C/10	83% 7 cycles	0.5 M Mg(ClO$_4$)$_2$ / AN	Pt metal	1 --1 vs. Ag/Ag$^+$
[164]$Mn_{0.4}V_2O_5$	144, 80	3, 2.2	0.05A/g, 2A/g	2 A/g: 82% 10,000 cycles	0.3 M Mg(TFSI)$_2$ / AN	Activated carbon	3.4-1.4
[166]τ-MnO$_2$	85	1.5-1.3			1 M Mg(ClO$_4$) / PC	Mg metal	1.9-1.3
[167]α-MnO$_2$	280	1.9-1	0.015 C	30% 6 cycles	0.2 M Mg-HMDS / THF	Mg metal	3-0.8
[168]δ-MnO$_2$	109	-0.5 --1.75 vs. Ag/Ag$^+$	0.1 A/g	50% 25 cycles	1 M Mg(ClO$_4$)$_2$ / AN	Mg metal	1 --2 vs. Ag/Ag$^+$
[170]Mn_3O_4	105,70	$\sim$1.1-0.6	0.1, 2 C	2 C: 94% 1,000 cycles	0.4 M 2:1 PhMgCl – AlCl$_3$	Mg metal	2.1-0.2
[172]$MgMn_2O_4$	120	0.4-0, -0.55 vs. Pt	10 µA	100% 24 cycles	0.5 M Mg(ClO$_4$)$_2$ / EC: DEC	V$_2$O$_5$	0.7 --0.6 vs. Pt
[173]$MgMn_2O_4$	250	$\sim$2.7, $\sim$1.7		28% 38 cycles	0.2 M Mg(TFSI)$_2$ / PC	carbon	3.6-1.6
[158]MoO_3	220	1.8	0.3 µA/cm^2		0.1 M Mg(TFSI)$_2$ / AN	Mg metal	2.8-1.7

Table 3.3 (cont.)

Material	Capacity (mAh/g)	Voltage	Rate	Retention	Electrolyte	Anode	Voltage Range
[177]MoS$_2$	170	2–1.7	0.02 A/g	95% 50 cycles	1:2 Bu$_2$ Mg – AlCl$_3$ / THF	Mg metal	3–0.5
[178]Ni[Fe(CN)$_6$]	40	~1.5–0.75 vs. carbon	10 mA/g	200% 50 cycles	1 M Mg(TFSI)$_2$ / PC	graphite	3 – –1 vs. Carbon

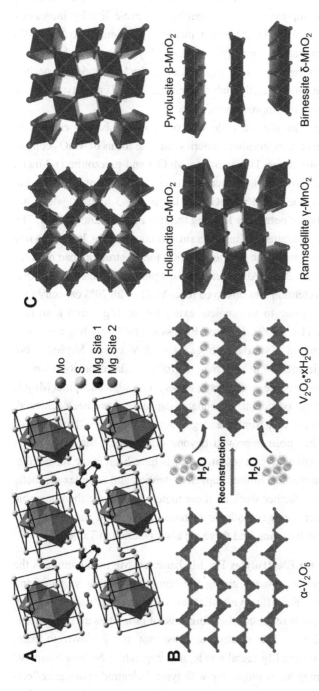

Figure 3.2 Crystal structures of (a) Mo_6S_8 reprinted with permission from ref [143], copyright Elsevier (2015), (b) V_2O_5 and hydrated $-V_2O_5$ reprinted with permission from ref [147], copyright CellPress (2019), and (c) various MnO_2 structures [148–151].

A

- ● Mo
- ● S
- ● Mg Site 1
- ● Mg Site 2

B

α-V_2O_5

$V_2O_5 \cdot xH_2O$

H_2O

Reconstruction

H_2O

C

Hollandite α-MnO_2

Pyrolusite β-MnO_2

Ramsdellite γ-MnO_2

Birnessite δ-MnO_2

of the cathode and passivation of the Mg anode [143–145, 162]. Incorporation of small amounts of metal ion, such as Mg or Mn, between V_2O_5 layers of the xerogels has led to improvements in capacity and cycle life by increasing structural integrity [163, 164]. Although many improvements to the V_2O_5 system have been made, the capacity is considerably lower than theoretical, and when hydrated xerogels are used, passivation of the Mg anode still occurs. For this system to be commercially viable, improvements in capacity and cycle life are key parameters for improvement.

Manganese oxide represents a wide variety of materials that can act as cathodes for rechargeable magnesium batteries. Several forms of MnO_2 contain tunnel and layered structures. Hollandite (α-MnO_2) and todorokite (τ-MnO_2) both contain tunnel structures created by edge-sharing MnO_6 octahedra. α-MnO_2 has 2×2 tunnels, and τ-MnO_2 has larger 3×3 tunnels, which are defined by the number of octahedra forming the side of each tunnel [165]. The large τ-MnO_2 tunnel structure only inserts a small amount of Mg, leading to low capacities [166]. Conversely, the smaller α-MnO_2 tunnel structure can produce much higher capacities >250 mAh/g, but had a low plateau voltage ~1.5 V [167, 168]. Poor capacity retention was observed for α-MnO_2 with 70% capacity fade in 6 cycles, which is due to incomplete extraction of Mg^{2+} and a surface conversion reaction [167, 169]. Layered birnessite (δ-MnO_2) has also been investigated, resulting in a similar voltage of ~1.5 V versus Mg/Mg^{2+} but much lower capacities of 109 mAh/g than α-MnO_2 similar capacity retention [168]. Other manganese oxide structures have been investigated, spinel Mn_3O_4 and spinel $MgMn_2O_4$ in particular. Reasonable capacities have been achieved, and with high-surface-area particles and graphene composites, cycle life of manganese spinels has been improved beyond other MnO_2 materials [170–174]. Overall, although extensive research efforts have been made for manganese oxide cathode materials, limited capacity retention and complex synthetic parameters necessitate further work on these materials to make them a viable system. Several other types of cathode materials have been investigated to a lesser extent for Mg batteries, and these are also included in Table 3.3.

Mg Metal Anode and Electrolytes For Mg batteries, Mg metal anode is the idealized choice for the system, giving the highest operating voltages and theoretical capacities. For effective utilization of Mg metal, selection of an appropriate electrolyte is required, therefore several properties and limitations need to be considered. The electrolyte must not break down to form a passivation layer on the Mg metal anode, and depending on how corrosive the electrolyte is it may be incompatible with typical electrode current collectors. For high-power applications, high capacities and voltages are needed;

therefore, an electrolyte with sufficient Mg ionic conductivity, and large voltage stability range is required. With these limitations in mind, the typical electrolytes used in Li-ion batteries cannot be used for a successful rechargeable Mg battery. Table 3.4 contains information about different electrolyte systems and their function in rechargeable magnesium batteries.

For many investigations of Mg cathode materials, acetonitrile or carbonate solvents with $Mg(ClO_4)_2$, $Mg(PF_6)_2$, or $Mg(TFSI)_2$ electrolyte salts are typically used; however, these electrolytes form thick passivation layers on Mg, causing an inhibition of Mg plating, severely diminishing cyclability and increasing Mg plating overpotential [179, 180]. To avoid passivation layer formation, organo-magnesium complexes were investigated. Mg organoborates and organo-aluminates in THF have shown good cyclability; however, only low capacities were achieved, and a maximum voltage of 1.5 V versus Mg/ Mg^{2+} could be used without oxidative electrolyte decomposition [180]. This led to ether-based solutions (tetrahydrofuran (THF) or glymes) containing magnesium halo-alkyl aluminate complexes (R_nAlCl_{3-n} Lewis acid with R_2 Mg Lewis base) [152]. By tuning the Lewis acid, electrolytes containing $RAlCl_2$, called dichloro complexes (DCCs), were produced, with the most successful being 0.25 M Bu_2 Mg with 0.5 M $EtAlCl_2$, which has an oxidative stability of 2.2 V versus Mg/Mg^{2+} and can achieve 100% coulombic efficiency for Mg platting/ stripping [152, 181–183]. To further push the oxidative stability limits, replacing all organic R groups with phenyl groups leads to all phenyl complexes (APCs), which can increase the oxidative stability over 3 V versus Mg/Mg^{2+} [175, 181, 184, 185]. An optimized electrolyte formulation for APCs is 0.5 M PhMgCl with 0.25 M $AlCl_3$, resulting in oxidative stability to 3.3 V versus Mg/ Mg^{2+}, 100% anodic coulombic efficiency, and < 0.2 V Mg plating overpotential [175].

The magnesium halo-alkyl aluminate complexes have drawn criticism due to the practicality of their complex synthesis requirements. To address this, fully inorganic Mg electrolyte salts were investigated and are termed magnesium aluminum chloride complexes (MACCs): $MgCl_2$-$AlCl_3$, $MgCl_2$-$AlPh_3$, $MgCl_2$-$AlEtCl_2$, and so on [186–188]. MAACs are used in a solvent of THF or glyme and have high oxidative stabilities of up to 3.4 V with improved reversibility with respect to the magnesiation of the cathode material [186]. By excluding organo-magnesium compounds, the production cost of these types of electrolytes are significantly less, which is highly advantageous when considering the scale of grid energy storage.

All the electrolytes just discussed, however, contain chloride, which can be quite corrosive, particularly to the common current collector metals: Al, Cu, Ni, stainless steel [189] when in the presence of water. As such, when using these

Table 3.4 A collection of various reports on electrolytes for rechargeable magnesium batteries. Included is the electrolyte salt and corresponding solvent, the oxidative stability (V vs. Mg/Mg^{2+}), the Mg plating overpotential (V), and the Mg plating coulombic efficiency.

Electrolyte Salt	Electrolyte Solvent	Oxidative Stability	Mg Plating Overpotential	Mg Platting CE
[180]0.25 M $Mg[B(Bu_2Ph_2)]_2$	THF/DME	1.50		99.6%
[152, 182]1:2 Bu_2 Mg – $AlCl_2Et$	THF	2	~0.26	95%
[181, 183]1:2 Bu_2 Mg – $AlCl_2Et$	THF	2.10	~0.4	75%
[181, 183]1:2 Bu_2 Mg – $AlCl_3$	THF	2.40	~0.55	86%
[181, 183]1:1.5 Bu_2 Mg – BPh_3	THF	1.77	~0.2	100%
[175]2:1 PhMgCl – $AlCl_3$	THF	3.3	0.195	98%
[184]1:1 PhMgCl – $Al(OPh)_3$	THF	>5	0.47	
[186, 187]2:1 $MgCl_2$ – $AlCl_3$	THF	3.4	0.27 – 0.35	90 – 100%
[190]0.5 M $Mg(BH_4)_2$	THF	1.7	0.6	40%
[190]0.1 M $Mg(BH_4)_2$	DME	1.7	0.34	67%
[191]0.1 M $Mg(BH_4)_2$ + 0.6 M $LiBH_4$	Diglyme	1.7	~0.25	99%

electrolytes, a different current collector is desired for long-term cycle stability; various carbon-based current collectors (glassy carbon, carbon cloth) are being considered. Alternatively, non–chloride-containing electrolytes have been investigated, which are compatible with standard current collector metals, most of which are boron based. Unfortunately, organo-borate, $Mg(BPh_2Bu_2)_2$, and borohydride, $Mg(BH_4)_2$, electrolytes both produce oxidative stability voltages less than 2 V versus Mg/Mg^{2+} [180, 190, 191].

Alloy Anode Materials Although Mg metal is the ideal anode for rechargeable magnesium batteries, the many complications associated with the Mg anode/electrolyte interactions have led researchers to investigate alloy-type anodes. These alloy anodes have potentials close to Mg/Mg^{2+} and can utilize traditional noncorrosive carbonate-based electrolytes [192]. Although better electrolyte compatibility can be obtained, there is an increase in anode material cost with most of the alloy metals and a significant decrease in theoretical capacity occurs. Many of the group IIIA, IVA, and VA metals have been investigated as a Mg alloy anode, but Bi and Sn are considered the two most promising and studied alloys. Table 3.5 contains information about different Mg alloy anodes and their function in rechargeable magnesium batteries.

Bismuth is promising due to its high Mg^{2+} diffusion coefficient and low insertion/extraction voltage (0.23/0.32 V vs. Mg/Mg^{2+}) [192, 193]. Upon magnesiation, Bi converts to Mg_3Bi_2, resulting in a theoretical capacity of 385 mAh/g. Although the theoretical capacity is low, through optimization of Bi morphologies, reasonable capacities, cycle retention, and high rate capabilities can be achieved [194–196]. Additionally, Bi has been shown to be compatible with traditional electrolytes, which is key for proper integration into a full cell with standard current collectors and high-voltage cathode materials [192, 197]. Tin has also drawn much interest as a Mg alloy anode because, similar to Bi, it has a low Mg diffusion barrier and insertion/extraction potential (0.15/0.20 V vs. Mg/Mg^{2+}) [193, 198]. Upon magnesiation, Sn forms Mg_2Sn which has a higher theoretical capacity of 900 mAh/g, well over double bismuth's theoretical capacity [192]. Unfortunately, poor rate capability and capacity retention is observed, so less than a third of the theoretical capacity is typically obtained, and the cycle life is inferior to that of Bi [192, 198, 199]. A unique Ga anode was reported, which has an exceptionally long cycle life of 1,000 cycles due to a self-healing effect from the low melting point of Ga metal (29.8°C) [200]. Gallium is capable of forming multiple different Mg-Ga alloys, so it was determined to form a Mg_2Ga_5 alloy, which has a theoretical capacity of 307 mAh/g. At 20°C the Ga anode failed to produce any current; however, at 40°C, 94% of theoretical capacity was achieved at 2 C after 680 cycles. This is

Table 3.5 A collection of various reports on alloy anode materials for rechargeable magnesium batteries. Included is capacity, voltage plateaus (V vs. Mg/Mg^{2+}), (dis)charging rate, capacity retention over the stated number of cycles, electrolyte used, and the voltage range at which the battery was cycled. All reports utilized Mg metal as a counter electrode.

Material	Capacity (mAh/g)	Voltage	Rate	Retention	Electrolyte	Voltage Range
[196]Bi	247	0.25	1 C	89% 100 cycles	1:2 EtMgCl – AlClEt$_2$ / THF	0.6–0.02
[195]Bi	325	0.2	2 C	93% 150 cycles	0.2 M/2 M Mg(BH$_4$)$_2$/LiBH$_4$ / diglyme	0.63–0.02
[199]Sn	425	0.15	0.01 C	53% 10 cycles	1:2 EtMgCl – AlClEt$_2$ / THF	0.6–0.1
[205]Sn	321	0.16	0.15 C	90% 30 cycles	0.5 M PhMgCl / THF	0.6–0.01
[200]Ga (40°C)	300	0.12	3 C	69% 1,000 cycles	0.4 M 2:1 PhMgCl – AlCl$_3$ + 0.4 M LiCl$_2$ / THF	0.7–0.01
[196]Sb	60	0.32	1 C	26% 50 cycles	1:2 EtMgCl – AlClEt$_2$ / THF	0.6–0.02
[196]Bi$_{0.88}$ Sb$_{0.12}$	298	0.27–0.29, 0.23, 0.12	1 C	72% 100 cycles	1:2 EtMgCl – AlClEt$_2$ / THF	0.6–0.02

[201]SnSb	345	0.17	0.5 A/g	78% 200 cycles	2:1 PhMgCl – AlCl$_3$ / THF	0.8–0
[202]Bi$_6$Sn$_4$	412	0.23, 0.16	0.2 A/g	68% 200 cycles	0.4 M 2:1 PhMgCl – AlCl$_3$ / THF	0.8–0
[203]Bi$_3$Sn$_2$	367	0.23, 0.16	1 A/g	93% 200 cycles	0.4 M 2:1 PhMgCl – AlCl$_3$ / THF	0.8–0

evidence of the liquid/solid Ga/Mg_2Ga_5 self-healing properties observed in this anode material.

Bimetallic anodes have also been investigated, showing interesting properties. Antimony by itself has a high theoretical capacity around 660 mAh/g; however, it shows extensive, irreversible Mg incorporation, making Sb a very poor alloy anode [196]. Due, however, to the high theoretical capacity of Sb, both $Bi_{1-x}Sb_x$ and $Sn_{1-x}Sb_x$ bimetallic anodes have been investigated. $Bi_{0.88}Sb_{0.12}$ demonstrated both improved capacity and capacity retention with respect to plain Bi [196]. The SnSb alloy anode showed a considerable enhancement in initial capacity of 150 mAh/g over Sn anode, had improved rate capabilities, and maintained 70% of its initial capacity after 200 cycles [201]. Arguably the most impressive bimetallic anode is that of bismuth and tin [202–204]. Bi_3Sn_2 produces higher capacities than either Bi or Sn alloys and has improved capacity retention of over 90% [202–204]. Further improvements in cycle life are still necessary for alloy anodes; however, the compatibility with conventional, noncorrosive electrolytes is promising for these types of anodes.

3.2.3 Current Status and Challenges

Currently, there are no reported applications of grid-scale energy storage based around the rechargeable magnesium battery system. Although many different types of cathode materials have been tested for the Mg battery, none contain all the necessary requirements of high capacity, cycle life, and rate capability. The overpotential, depth of discharge, Coulombic efficiency for Mg platting/stripping, and anode current collector selection will all require further progress prior to incorporation into a full cell. Development of a wide variety of electrolytes has been very beneficial in recent years, and the differences in properties, benefits, and limitations of each allow for informed combinations with cathode materials. The high theoretical gravimetric and volumetric capacity, along with cost and improved safety associated with Mg batteries warrant further investigation; however, there are several fundamental science challenges yet to be solved in order to make this system ready for grid-scale energy storage.

3.3 Rechargeable Aqueous Zinc Batteries (RAZB)

3.3.1 Overview

Similar to the rechargeable Mg battery system, rechargeable Zn batteries have the benefit of utilizing a divalent ion. Due to the significant increase in density and weight of Zn metal, however, much lower theoretical gravimetric capacities are obtained, only 820 mAh/g [206–208]. Additionally, the Zn/Zn^{2+} reduction potential is -0.76 V versus SHE, which is very positive compared to Li, Na, and Mg [206–208]. These seemingly negative attributes of Zn are also its best properties. Although the gravimetric capacity is low, the theoretical volumetric capacity is 5,857 mAh/cm^3, which is the second highest volumetric capacity of all the standard battery systems (Al: 8,046 mAh/cm^3) [94, 206–208]. Since the reduction potential is more positive than other battery systems, water can be used as an electrolyte, as the water reduction potential can be reasonably suppressed beyond -0.76 V versus SHE, allowing for a Zn metal anode in aqueous electrolyte. While Zn is less abundant than Mg, due to easy industrial processing and well-established commercial availability, Zn is very inexpensive, helping bring the overall cell cost down. There are many desirable properties of the aqueous zinc battery system; however, complications still plague the system in different ways. At the anode, dendritic zinc growth occurs, which will short the battery upon extended cycling, and H_2 evolution induces side reactions that corrode the Zn anode [206, 207, 209]. Due to the high electrostatic charge of Zn^{2+}, it is challenging to determine a suitable cathode material that is capable of incorporating Zn^{2+} and doing so without causing dissolution products and irreversible structural changes [209–211]. If these complications can be addressed, the high volumetric capacity, low cost of Zn and aqueous electrolytes, high safety, ease of assembly, and low toxicity make the rechargeable aqueous zinc battery an extremely promising battery system for grid-scale energy storage. Batteries that utilize the alkaline electrolyte are discussed in detail in another section in this Element, therefore this section will focus only on the mildly acidic Zn electrolyte.

3.3.2 Cell Components

Cathode Materials Similar to the rechargeable magnesium battery, the increased electrostatic interactions of the divalent zinc ion cause challenges in the reversible interactions of Zn^{2+} and cathode materials [206, 207, 209]. When utilizing an aqueous electrolyte, additional cathode considerations must be made. A high plateau voltage is desired to gain a larger operating voltage; however, the voltage must not allow for water oxidation to occur as a competing reaction on the cathode surface. The mildly acidic conditions of the typical Zn electrolyte can also be detrimental to cathode materials during electrochemical battery operation [206, 207, 209]. To address these initial requirements, many materials have been investigated; however, manganese oxide and vanadium oxide type materials have stood out as the most promising candidates for the reversible incorporation of Zn^{2+} [206–210].

Manganese Oxide MnO_2 was previously discussed as a rechargeable magnesium cathode; however, the interactions of Zn^{2+} in aqueous electrolyte lead to very different chemistries than those of Mg^{2+} in nonaqueous electrolyte. Many of the various MnO_2 phases (α^-, β^-, γ^-, τ^-, and δ^-MnO_2) have been tested for Zn batteries, and MnO_2 crystal structures are presented in Figure 3.2 in the RMB section [206–210]. At this time, there is still not a well-agreed-upon mechanism for how Zn^{2+} interacts with MnO_2, and there is evidence to support three different reactions [206, 210]. The first is a standard (de)intercalation reaction where Zn^{2+} is inserted to form spinel $ZnMn_2O_4$, layered Zn-buserite, or layered Zn-birnessite [212–215]. A conversion reaction was also proposed where MnOOH is formed by removal of H^+ from water and the subsequently generated OH^- causes precipitation of a $ZnSO_4[Zn(OH)_2]_3 \cdot xH_2O$ phase on the surface of the cathode [216]. A two-ion co-insertion reaction was also proposed where H^+ is initially inserted with rapid kinetics followed by a slower Zn^{2+} insertion, finally forming spinel $ZnMn_2O_4$ [217]. It has not yet been confirmed which of these mechanisms is most likely to occur under specific operating conditions, or if certain MnO_2 phases will preferentially follow one or more of the mechanisms.

Table 3.6 includes data for several selected examples of manganese oxide cathode materials and their performance for use in a RAZB. Of these, Hollandite α-MnO_2 is one of the most commonly studied materials and has a MnO_6 octahedron based 2×2 tunnel structure, which gives ample room for Zn^{2+} insertion. All three mechanisms have been observed for α-MnO_2, but all have average voltage plateaus within the range of 1.3–1.4 V versus Zn/Zn^{2+} [215–217]. Capacities of 200–300 mAh/g can be achieved, and thousands of cycles with reasonable retention at high rates have been reported [215–217]. Due to the narrow 1×1 tunnels in pyrolusite β-MnO_2, it is expected that Zn^{2+} will have difficulty inserting with reasonable kinetics [214]. Through the use of crystallographic, imaging, and synchrotron techniques, it was determined that upon the first insertion of Zn into β-MnO_2, a structural change occurs, forming the layered structure Zn-buserite, which does not revert back to β-MnO_2 upon deinsertion [218]. This permanent structural change results in reasonable capacities, cycle life, and rate capabilities [218, 219]. Ramsdellite, γ-MnO_2, has narrow 1×1 tunnels, but also contains 1×2 tunnels that can allow for easier Zn^{2+} insertion compared to β-MnO_2. Unfortunately, a conversion to spinel $ZnMn_2O_4$, tunnel Zn_xMnO_2, and layered Zn_xMnO_2 occurs, and only some of the new phases revert back to γ-MnO_2 upon charging, resulting in significant capacity fade [220]. Todorokite τ-MnO_2, has large 3×3 tunnels (7 Å), which can easily accommodate Zn^{2+} intercalation; however, minimal attempts have been made to utilize this material, and only low capacities were achieved [221]. Unlike the other MnO_2 materials, which contain tunnel-like structures, birnessite δ-MnO_2 is

Table 3.6 A collection of various reports on manganese oxide cathode materials for RAZB. Included is capacity at a low and a high rate (where applicable), approximate voltage plateau (V vs. Zn/Zn^{2+}), (dis)charging rate associated with the two capacities, capacity retention over the stated number of cycles, electrolyte salt used, and the voltage range over which the battery was cycled at. All reports utilized a Zn metal anode for the battery configuration.

Material	Capacity (mAh/g)	Voltage	Rate	Retention	Electrolyte Salt	Voltage Range
[215]α-MnO$_2$	210, 130	~1.4–1.3	C/2, 6 C	6 C: 73% 100 cycles	1 M ZnSO$_4$	1.9–1
[216]α-MnO$_2$	260, 161	1.4, 1.3	1 C, 5 C	5 C: 92% 5,000 cycles	2 M ZnSO$_4$, 0.1 M MnSO$_4$	1.9–1
[217]α-MnO$_2$	290, 70	1.4, 1.3	0.3 C, 6.5 C	6.5 C: 78% 10,000 cycles	2 M ZnSO$_4$, 0.2 M MnSO$_4$	1.8–1
[219]β-MnO$_2$	270, 180	1.4, 1.3	0.1, 0.2 A/g	0.2 A/g: 75% 200 cycles	1 M ZnSO$_4$, 0.1 M MnSO$_4$	1.8–1
[218]β-MnO$_2$	258, 151	1.3, 1.2	0.65 C, 6.5 C	6.5 C: 94% 2,000 cycles	3 M Zn(CF$_3$SO$_3$)$_2$, 0.1 M Mn(CF$_3$SO$_3$)$_2$	1.9–0.8
[220]γ-MnO$_2$	230	1.4, 1.3	0.5 mA/cm^2	~63% 45 cycles	1 M ZnSO$_4$	1.8–1
[221]τ-MnO$_2$	108	1.4, 1.25	C/2	~100% 50 cycles	1 M ZnSO$_4$	2–0.7
[226]δ-MnO$_2$	250	1.4, 1.3	83 mA/g	~45% 100 cycles	1 M ZnSO$_4$	1.8–1

Table 3.6 (cont.)

Material	Capacity (mAh/g)	Voltage	Rate	Retention	Electrolyte Salt	Voltage Range
[222]δ-MnO$_2$	238, 138	1.5–1.2	C/5, 20 C	20 C: 93% 4,000 cycles	1 M Zn(TFSI)$_2$, 0.1 M Mn(TFSI)$_2$	1.8–1
[223]δ-MnO$_2$	350, 154	1.5, 1.4–1.2	0.1, 3 A/g	3 A/g: 75% 200 cycles	1 M ZnSO$_4$	1.9–1
[224]δ-MnO$_2$	260, 100	1.4, 1.3	0.3, 2 A/g	2 A/g: 100% 5,000 cycles	2 M ZnSO$_4$, 0.1 M MnSO$_4$	1.8–1
[225]δ-MnO$_2$	320, 155	1.6–1.1	1 C, 10 C	10 C: ~60% 1,000 cycles	0.25 M ZnSO$_4$, 0.75 M Na$_2$SO$_4$	1.9–0.9
[227]ZnMn$_2$O$_4$	125, 90	1.4	0.1, 0.5 A/g	0.5 A/g: 94% 500 cycles	3 M Zn(CF$_3$SO$_3$)$_2$	2–0.8
[228]K$_{0.8}$Mn$_8$O$_{16}$	325, 150	1.4, 1.3	0.1, 1 A/g	1 A/g: 100% 1,000 cycles	2 M ZnSO$_4$, 0.1 M MnSO$_4$	1.8–0.8

a layered structure with a ~ 7 Å interlayer spacing. This layered structure allows for facile (de)intercalation of Zn^{2+} and can achieve capacities, cycle life, and rate capability comparable to those of α-MnO_2 [222–225].

Vanadium Oxide Similar to the rechargeable Mg battery cathode, V_2O_5 can also accommodate the high electrostatic charge of the Zn^{2+} ion for intercalation within its layered structure. Table 3.7 includes data for several selected examples of vanadium oxide cathode materials and their performance for use in a rechargeable aqueous zinc battery. Pristine V_2O_5 has low ionic and electrical conductivity, as well as suffers from structural instability, leading to generally lower capacities, capacity retention, and rate capabilities; however, through modification of the V_2O_5 structure, drastic improvements can be made [229, 230]. Hydration of V_2O_5 causes a structural change from a single to a bilayer structure, which provides large interlayer spacings of > 10 Å, resulting in improved capacity, capacity retention, and rate capabilities [231]. The improvements upon hydration were correlated with both the increased interlayer spacing and a shielding effect of the Zn^{2+} electrostatic charge by the neutral water molecules. Even more impressive improvements can be made by introduction of interlayer metal ions (Zn, Na, Ca, Mg, Li, etc.), which also form a bilayer structure that contains MO_6 octahedra that pillar the V_2O_5 bilayers, increasing interlayer spacing to > 10 Å and improving structural rigidity [232–235]. This enables drastically improved capacities, cycle life, and rate capabilities, generating in the case of $Li_xV_2O_5$, 450 mAh/g at 0.5 A/g and ~87% retention after 1,000 cycles at 10 A/g [235].In addition to the modified V_2O_5 materials, other vanadium oxide compounds such as V_6O_{13}, $Na_2V_6O_{16}$•$1.63H_2O$, $H_2V_3O_8$, and more have also been tested as aqueous Zn metal battery cathodes [229, 236, 237]. Functional capacity, voltage, and capacity retention of these other vanadium oxide compounds can be seen in Table 3.7. One major challenge for the various vanadium oxide cathodes is they all possess low sloping plateau voltages with an average typically in the range of 0.7–0.8 V versus Zn/Zn^{2+}, which is significantly lower than the MnO_2 compounds, which all have average voltage plateaus around 1.3 V versus Zn/Zn^{2+}. This lower voltage will detrimentally affect the specific energy of the full cell battery.

Zn Metal Anode The use of a Zn metal anode for the rechargeable aqueous zinc battery is key because it allows access to the high volumetric capacities of Zn. To effectively use a Zn metal anode, a suppression of dendritic growth, high coulombic efficiency for platting/stripping Zn metal, and minimal H_2 evolution and side reactions are required [206–209]. To address these common complications three main strategies have been developed: structural design of the Zn electrode, Zn surface modification, and electrolyte optimization.

Table 3.7 A collection of various reports on vanadium oxide cathode materials for RAZB. Included is capacity at a low and a high rate (where applicable), approximate voltage plateau (V vs. Zn/Zn^{2+}), (dis)charging rate associated with the two capacities, capacity retention over the stated number of cycles, electrolyte salt used, and the voltage range the battery was cycled at. All reports utilized a Zn metal anode for the battery configuration.

Material	Capacity (mAh/g)	Voltage	Rate	Retention	Electrolyte Salt	Voltage Range
[230]V_2O_5	224, 113	1, 0.6	0.1, 2 A/g	2 A/g: ~100% 400 cycles	3 M $ZnSO_4$	1.4–0.4
[231]$V_2O_5 \cdot nH_2O$	372, 300	1–0.9, 0.6–0.4	0.3, 6 A/g	6 A/g: 71% 900 cycles	3 M $Zn(CF_3SO_3)_2$	1.6–0.2
[232]$Zn_{0.25}V_2O_5 \cdot nH_2O$	300, 260	1, 0.8, 0.7	1 C, 8 C	8 C: 80% 1,000 cycles	1 M $ZnSO_4$	1.4–0.5
[234]$Na_{0.33}V_2O_5$	367, 218	0.9–0.5	0.1, 1 A/g	1 A/g: 93% 1,000 cycles	3 M $Zn(CF_3SO_3)_2$	1.6–0.2
[233]$Ca_{0.25}V_2O_5 \cdot nH_2O$	340, 72	1–0.9, 0.8, 0.7	C/2, 80 C	80 C: 76% 5,000 cycles	1 M $ZnSO_4$	1.6–0.6
[235]$Li_xV_2O_5 \cdot nH_2O$	386, 192	1.05–0.95, 0.7–0.5	1, 10 A/g	10 A/g: ~85% 1,000 cycles	2 M $ZnSO_4$	1.4–0.4
[229]V_6O_{13}	450, 206	~1–0.5	0.1, 10 A/g	10 A/g: ~100% 3,000 cycles	3 M $ZnSO_4$	1.4–0.3

[235]Na$_2$V$_6$O$_{16}$·1.63H$_2$O	296, 155	~1–0.4	0.1, 5 A/g	5 A/g: 90% 6,000 cycles	3 M Zn(CF$_3$SO$_3$)$_2$	1.6–0.2
[238]NaV$_3$O$_8$·1.5H$_2$O	375, 150	0.9–0.8, 0.6–0.5	0.1, 4 A/g	4 A/g: 83% 1,000 cycles	1 M ZnSO$_4$, 1 M Na$_2$SO$_4$	1.3–0.3
[236]H$_2$V$_3$O$_8$	428, 150	0.8, 0.6–0.5	0.1, 5 A/g	5 A/g: 94% 1,000 cycles	3 M Zn(CF$_3$SO$_3$)$_2$	1.6–0.2

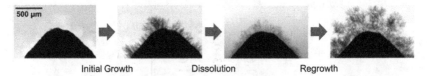

Initial Growth   Dissolution   Regrowth

Figure 3.3 Zn dendritic growth, dissolution, and regrowth on a Zn metal cone in an alkaline solution. Adapted with permission from ref [239], copyright CellPress (2019).

The dendritic growth of Zn is exceptionally detrimental because secondary dendrites can branch off from the parent dendrites, extending the growths until the battery separator is penetrated and the battery is shorted (Figure 3.3) [209, 239, 240]. Additionally, dendrites can mechanically break away from the electrode, leading to "dead" or "orphaned" zinc, which lowers coulombic efficiency and causes capacity fade [209, 239, 240]. The use of 3D structures has been found to effectively inhibit dendrite formation due to the uniform distribution of charge across its surface and the structural limitations of interior dendritic growth [206, 241, 242]. The use of 3D Zn foam, 3D porous copper template, imbedding Zn nanoparticles in a carbon matrix, and Zn deposited on high-surface-area carbon fiber, carbon cloth, and carbon nanotubes have all been demonstrated to greatly improve depth of discharge, platting/stripping efficiency, cycle life, and suppression of dendritic growth [243–248].

Surface coatings have been used to target decreased dendritic growth as well as prevent hydrogen evolution, which generates zinc oxide and hydroxide-based corrosion products that block and passivate the Zn surface. Various polymer coatings can be used to protect the surface while shuttling the Zn ions through the polymer layer. Polymer coating is an extremely effective method for inhibition of dendritic growth and corrosion caused by H_2 generation; however, a considerable increase in Zn platting/stripping overpotential occurs due to the insulating surface layer [249]. To mitigate the high overpotentials of polymer coatings, various thin-film or porous inorganic coatings have been utilized, such as 8 nm TiO_2, and porous $CaCO_3$ and ZnO_2 [250–252]. These inorganic coatings can suppress the dendritic Zn growth and H_2 evolution as effectively as can the polymer coatings but retain only minimal increases in Zn platting/stripping overpotential.

Mildly Acidic Aqueous Electrolyte Due to the more positive redox potential of Zn/Zn^{2+}, either organic or aqueous electrolytes can be utilized with rechargeable zinc batteries. The higher conductivity, lower toxicity and cost, and inherent nonflammability and safety of aqueous electrolytes make aqueous electrolytes highly attractive for grid-scale energy storage. Typically, aqueous zinc battery electrolytes fall into two categories: alkaline electrolyte, which utilizes concentrated basic solutions, and mildly acidic electrolyte, which has a pH of about 4, depending on the electrolyte salt and concentration. Several common zinc salts have been tested; however, zinc sulfate ($ZnSO_4$) and zinc trifluoromethanesulonate ($Zn(CF_3SO_3)_2$) are widely accepted to be the best electrolyte salts for standard aqueous electrolytes [227, 230]. Higher concentrations of $Zn(CF_3SO_3)_2$ can be achieved, reaching 3 M electrolyte compared to 2 M $ZnSO_4$, providing the dual benefits of improved electrical conductivity and decreased solvation effect of H_2O on Zn since the bulky $CF_3SO_3^-$ ligands limit the number of H_2O molecules around the Zn ion [227, 253]. This leads to improved Zn charge transfer to the cathode materials and improved metal platting/stripping, thereby decreasing the dendritic growth of Zn. Although $Zn(CF_3SO_3)_2$ has typically shown better performance than $ZnSO_4$, the increased cost of the $Zn(CF_3SO_3)_2$ salt has led to $ZnSO_4$ being the primary interest for practical aqueous zinc batteries [206, 253].

In addition to the selection of the electrolyte salt, many additives have been tested to improve cycle life of aqueous zinc batteries. Several additives like $MnSO_4$ and Na_2SO_4 have been shown to help enhance the stability of particular cathode materials by preventing cathode dissolution, and Na_2SO_4 also was shown to help avoid the formation of Zn metal dendrites [238]. It is believed that the Na^+ present in solution will be attracted to the areas of high electrostatic charge where dendrite formation occurs and will create a positively charged "shielding" effect that prevents Zn^{2+} from depositing and forming dendrites at those locations [238, 254]. Polymeric additives can be used in the electrolyte to control the distribution of current near dendrite tips, causing a more uniform charge distribution across the surface of the zinc [240, 255, 256]. Various organic surfactants and capping agents can have similar effects to those of the polymeric additives by adsorbing to the electrode surface; however, the discrete organic additives also affect H_2 evolution and thereby suppress corrosion products [257, 258]. Moving forward, the electrolyte of choice for aqueous Zn batteries will likely be 2 M $ZnSO_4$ in aqueous media with the addition of cost-effective additives specific to the cathode material and anode configuration in the battery.

3.3.3 Challenges

Currently, there are several key complications that occur when using manganese oxide or vanadium oxide cathode materials. For MnO_2, manganese dissolution occurs during cycling due to structural changes during phase transitions, and disproportionation of Mn^{3+} ions [259, 260]. One of the more straightforward methods to avoid the dissolution of Mn into solution is by intentionally adding 0.1 M Mn^{2+} to the electrolyte solution to change the equilibrium dynamics of the Mn/Mn^{2+} dissolution [216, 261]. Structural modifications of the MnO_2 cathode itself have also shown promising results, where incorporation of K^+ ion into the tunnels of α-MnO_2 stabilizes the Mn polyhedra and creates oxygen defects that can improve Zn^{2+} diffusion and electrical conductivity [207, 228]. Similar to MnO_2, vanadium dissolution from vanadium oxide cathodes is caused by various soluble V redox inter-mediate compounds, which lead to low capacity retention [207, 211]. In addition to loss of active cathode material, dissolved vanadium compounds deposit on the Zn metal anode, causing passivation. V dissolution is actually a slow process, however, and the amount of dissolution and capacity fade at high rates of (dis)charge is less than at slower rates [207, 211, 238]. Unlike Mn dissolution, there are no well-established methods for mitigating V dissolution. Encapsulation of the cathode in conductive carbon can min-imize dissolution by limiting direct cathode interaction with the aqueous electrolyte [211, 231]. In addition to complications derived from V dissolution, irreversible phase transitions have been observed with the formation of $Zn_3V_2O_7$. In this compound, Zn forms a stable ZnO_6 octahe-dron, from which it is difficult to extract Zn^{2+} upon deintercalation [262, 263]. Over repeated cycling, buildup of $Zn_3V_2O_7$ leads to capacity fade.

3.3.4 Current Status

By utilizing a Zn metal anode in an aqueous electrolyte, significant improve-ments in volumetric capacity, cost, and safety can be made with respect to current battery systems. Urban Electric Power utilizes the same chemistry as disposable alkaline batteries, producing Zn–MnO_2 batteries in alkaline electro-lyte. Urban Electric Power primarily makes smaller battery units for residential use, but is currently developing a 1 MWh system for the City University of New York and City College of New York [11, 13]. Currently there are no grid-scale deployment projects for the rechargeable aqueous zinc battery using a mildly acidic electrolyte. For the mildly acidic electrolyte batteries, the 2 M $ZnSO_4$ aqueous electrolyte has already been established as an effective low-cost electrolyte that can be paired with a variety of additives to help prevent H_2 evolution and Zn corrosion, cathode dissolution, and Zn dendrite formation. Both manganese and vanadium oxide cathodes show promise, each with their specific advantages. With further optimization of these two cathode materials, incorporation into a fully functional rechargeable aqueous Zn battery system could be used for energy storage at grid scale.

3.4 Metal–Air Batteries

3.4.1 Overview

Metal–air batteries, which contain a metal anode and an air-breathing cathode, have theoretical energy densities much higher than those of lithium-ion batteries. Thus, metal–air batteries are frequently advocated as promising candidates toward next-generation electrochemical energy storage for applications including electric vehicles and grid energy storage [264]. The metal anode can include alkali metals (Li, Na, and K), alkaline earth metals (Mg), or first-row transition metals (Fe and Zn), and the electrolyte can be aqueous or nonaqueous. However, currently there is no operational deployment of such systems in large-scale GEES, due to challenges associated with the metal anode, air cathode, and electrolyte. The Zn–air battery in an aqueous system will be chosen as a model system in this section, as it represents the most promising candidate for GEES deployment.

3.4.2 Zn–Air Battery

The Zn–air battery is a century-old technology. The concept was first reported by Smee in 1840, and the first commercial rechargeable Zn–air batteries were developed by NantEnergy in 2012, which was reported to have a limited energy density of ~35 Wh/kg in 2017 [265]. Today, the Zn–air battery company NantEnergy has deployed over 120 microgrids in Indonesia, Africa, and even in the US (working with Duke Energy) that are entirely powered by solar panels and energy storage with no fossil fuels. The Zn–air battery has a theoretical energy density of 1,353 Wh/kg (excluding oxygen) with a theoretical working voltage of 1.65 V and very low fabrication cost [28].

Reaction Mechanism The Zn–air battery combines fuel cell technology with traditional solid-state electrode design. During discharge, Zn is oxidized to Zn^{2+} at the anode, while O_2 from the surrounding air is reduced to form OH^- on catalyst particles supported at the gas-diffusion cathode (Figure 3.4). These reactions may be reversed upon charge with Zn plated at the anode and O_2 evolving at the cathode [28]. The oxygen reduction reaction (ORR) taking place at the cathode side is similar to that in hydrogen fuel cells, thus sharing several common design principles and criteria. Zn–air batteries can be recharged either mechanically by replenishing the Zn anode and electrolyte, or electrically using bifunctional oxygen electrocatalysts or a combination of the ORR and OER catalysts.

Air Cathode Air cathodes are constructed by loading electrocatalysts onto porous polytetrafluoroethylene (PTFE) treated gas diffusion layers, which have

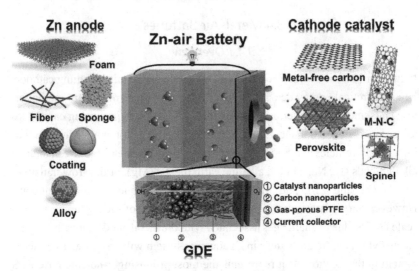

Figure 3.4 Schematic of Zn–air batteries including the gas-diffusion electrode (GDE) structure, different candidate materials for cathode electrocatalysts, and different forms of Zn anode materials. Figure reprinted from ref [265]. Published by The Royal Society of Chemistry.

balanced hydrophobic and hydrophilic properties [266]. They allow fast permeation of O_2 and provide abundant gas–electrolyte–electrode interfaces for ORR to happen at a high rate. Electrocatalysts are the critical components responsible for accelerating ORR and OER, and thereby determine the power density, energy efficiency, and cycle life of the battery. Pt represents the benchmark catalyst for ORR, while IrO_2 and RuO_2 are the most efficient catalysts for OER [267]. However, due to their prohibitive costs, conventional Zn–air batteries often use MnO_2 as the bifunctional cathode electrocatalyst [268], the limited activity and stability of which result in the poor power density of conventional Zn–air batteries. Extensive research efforts have been spent in the search for better cathode electrocatalysts. Various non–precious-metal-based candidates such as metal-free carbonaceous materials, M-N-C materials, as well as transition metal oxides and sulfides have been identified as the necessary materials for Zn–air battery application. The lower intrinsic activities of these catalysts can be compensated by increasing areal loading to achieve similar or even superior activity to that of Pt in a concentrated alkaline solution. Remarkable peak power density (up to >400 mW/cm^2) has been reported using $La_{0.99}MnO_{3.03}$/C electrocatalysts at room temperature [136]. In terms of electrically recharged Zn–air batteries, using state-of-the-art cathode electrocatalysts, due to the large ORR and OER overpotentials, only low round-trip energy efficiency of <65% under real working conditions are achieved [265]. This represents an intrinsic drawback of rechargeable Zn–air batteries when compared to conventional lithium-ion batteries, denoting an energy efficiency of 80–90%. In addition, the harsh, concentrated alkaline solution is detrimental to the ORR-active component due to corrosion of the carbon support and leaching of transition metals [28].

Zn Anode The Zn anode plays an important role in Zn–air batteries, as the parasitic reaction between Zn and the electrolyte can lead to spontaneous H_2 generation and electrode corrosion, lowering the utilization of active materials. The corrosion tends to be further aggravated in concentrated alkaline electrolyte solutions, due to the lack of surface oxide passivation layers on the Zn anode. Furthermore, dendritic Zn tends to form upon charging when exposed to nonuniform distribution of current density at the electrode surface, which may penetrate the separator, leading to a short circuit and failure in the battery. Compared to the intense research efforts on cathode electrocatalysts, there are far fewer reports on Zn anodes due to the excessive amounts of Zn metals and electrolytes, as well as the low depth of discharge used in previous air cathodes-related studies, which barely challenge Zn anodes [269]. However, practical applications at large scales require high Zn utilization and reversibility under realistic conditions.

Currently, several strategies to improve Zn anode performance have been reported, including preparations of high-surface-area Zn anodes, such as fibers, sponges, and foams, which enhance electrochemical performance, but often come at the price of faster electrode corrosion due to enlarged contact area with the electrolyte. Additionally, the use of compositionally modified Zn anodes through alloying with certain metals (Pb, Cd, Bi, Sn, and In) can suppress H_2 evolution, increase conductivity, and improve current distribution or implementation of surface trapping layers (TiO_2, TiN_xO_y) to increase electrode reversibility. Despite the various research efforts, challenges still exist as the cycling stability of Zn anodes, especially under deep discharge conditions, still falls short of expectations, with only a few examples demonstrating fully cyclable Zn anodes for more than 100 cycles [270].

Electrolytes Electrolytes are critical components for Zn–air batteries and searching for a stable, low-volatility, non-toxic, and high-oxygen-solubility electrolyte with a wide electrochemical window is a common goal for the metal–air battery systems. Current state-of-the-art Zn–air batteries utilize an alkaline electrolyte like KOH for its high conductivity, good electrochemical reaction kinetics, and moderate Zn solubility [271]. However, the absorption of CO_2 from air into the electrolyte leads to formation of parasitic CO_3^{2-} species, which poisons the electrolyte and results in a shortened lifetime of alkaline Zn–

air batteries when exposed to air. Non-alkaline electrolytes have been recently probed as alternatives to improve the Zn cyclability. The $ZnCl_2$–NH_4Cl near-neutral aqueous electrolyte has been reported to suppress the Zn corrosion and dendrite growth, but led to a less active Zn anode and slower ORR and OER [272]. Other nonaqueous electrolytes such as ionic liquids are beneficial for Zn stability but completely alter the reaction pathways of ORR and OER and drastically reduce the reaction kinetics. Hence, the nonaqueous electrolytes are not very suitable for Zn–air batteries, especially when high powder density is heavily demanded [265].

3.4.3 Current Status and Challenges of Zn–Air Batteries

Owing to the intense research efforts over the past decade, the performance of Zn–air batteries is no longer limited by air cathodes and electrocatalysts. It is believed that a quick shift of research focus from air cathodes to Zn anodes would therefore greatly benefit this community and eventually realize the full potential of this century-old technology [265]. To date, NantEnergy has 3,000 energy storage systems installed in nine different countries with over six years of deployment and has recently dispatched over 6 GWh of clean energy. Recently, New York State has just signed a deal with Canadian company Zinc8 Energy Storage to procure a new 100 kW/1 MWh Zn–air battery, which will be installed behind-the-meter in an undetermined western New York site that can benefit from such hardware [273]. It is essential for the rapid transformation of Zn–air technology from mainly laboratory-scale science to large-scale applications. In the next ten years, it has been suggested that an achievable goal would be to develop Zn anodes with high active material utilization (>80%), capable of deep discharge and charge (DOD > 50%), with reasonable cycle life (>500 cycles) and high coulombic efficiency (>80%) [265].

3.4.4 Other Metal–Air Batteries

Several other metal–air batteries have been reported in aqueous media, such as Fe–air, Al–air, and Mg–air batteries. Fe–air batteries can be electrically rechargeable, but their low practical energy density of 60–80 Wh/kg is far less than current lithium-ion technology. However, they can be suitable for stationary energy storage owing to their long cycle life (>1,000 cycles), low cost (< $100/kWh), and environmentally benign character [274]. The Al–air and Mg–air batteries have significant theoretical energy density; however, only a limited fraction can be utilized due to the severe parasitic corrosion of the metal anodes in contact with aqueous electrolyte, which renders them only mechanically rechargeable [266]. Nonaqueous metal (Li, Na, K)-air batteries utilize aprotic electrolytes, which lead to ORR with drastically different mechanisms, forming peroxide MO_2 [28]. These superoxides or peroxides have limited solubility in the electrolyte and have a tendency to deposit on the air cathode, resulting in the gradual blockage of the available cathode surface area and eventually shutting off the battery. Therefore, the discharge capacity of nonaqueous metal–air is far smaller than the theoretical value. Compared to aqueous metal–air batteries, nonaqueous metal–air batteries are still in their infancy. They suffer from even more serious challenges, and therefore will not be discussed in detail in this section.

4 Conclusion

This Element introduced several types of potential grid-scale electrochemical energy storage systems (GSEESSs) (lead–acid batteries, redox flow batteries, Na–S batteries, metal–air batteries, Na-ion batteries, as well as Mg and Zn batteries) that are beyond Li-ion battery technology. Details regarding battery mechanisms, compositions, performance metrics, advantages, and disadvantages were discussed to provide a road map to guide future research and development to promote the commercial application of GSEESSs.

In spite of the tremendous progress achieved to date, several challenges still exist before full deployment of these batteries as GSEESSs.

1. Cost-effective technology is still urgently needed. The use of Li-ion batteries in stationary applications is costly [275]. The current redox flow battery technologies, including VRBs, are still prohibitively expensive in terms of both capital and life-cycle cost. Further research and development is required in the fields of materials sourcing and improving cell stack power density.

2. System stability and safety are critical factors that demand further investigation for development of battery technologies for GSEESS. For RFBs containing Br_2 such as ZBB, the high cell voltage and highly oxidative Br_2 demand expensive cell electrodes, membranes, and flow-controlling components that have high chemical stability to mitigate for the high toxicity of Br_2 through inhalation and absorption and further maintain system stability. For the Zn-based batteries such as ZBB, RZAB, and Zn–air batteries, repeated plating of Zn in general can induce formation of dendrites that might penetrate the separator, thus leading to cell failures. Controlled operating modes, such as pulsed discharge during charge, are often required to achieve uniform plating and reliable operation. For systems containing sulfur such as PSBs, the system often encounters sulfur loss during extended

cycling due to the buildup of sulfur species (e.g. S^{2-} and/or HS^-) either on the electrodes or membrane. In addition, mixing of the electrolytes can generate heat and toxic gases such as Br_2 and H_2S, which raise safety concerns. For Na–S batteries, the high operating temperature range of the HT Na–S system raises several safety concerns. The reactive sodium needs to be stored in a corrosion-resistant safety tube equipped with a small supply hole that regulates the flow of sodium and also allows a minimum amount of sodium to maintain the electrochemical reaction. Sulfur is highly flammable and can generate toxic SO_2 gas upon oxidation. In addition, due to the hygroscopic nature of polysulfides a water-free environment is a prerequisite for stable system operation, which requires inner and outer protection sheets to be added for thermal insulation and fire resistance.

3. Current lead–acid batteries suffer from a limited depth of discharge, battery efficiency, and cycle life, while Na–S and flow batteries still face challenges in high-rate operations. Therefore, both fundamental and applied research in multiple areas are required, encompassing fundamental material chemistry studies to systems-level battery design, to further advance the current technologies.

4. The duration of discharge for many of the GSEESSs is limited to fewer than four hours, which is insufficient for load leveling throughout the entire peak power period. RFBs show promise in scalability; however, in general, further improvements to capacity are required for long-term load leveling (>10 hours).

As the requirements for GSEESS are very different from those for consumable electronics and transportation, the ultimate solution may come from a different chemistry. Continued exploration of low–technology, readiness-level energy storage systems that could provide high reward in terms of high safety, low cost, and/or long cycle life, such as those highlighted in Section 3 of this Element, will be essential to develop the best future GSEESS solutions.

References

[1] X. Fan, B. Liu, J. Liu *et al.*, "Battery Technologies for Grid-Level Large-Scale Electrical Energy Storage," *Transactions of Tianjin University*, vol. 26, pp. 92–103, 2020. https://doi.org/10.1007/s12209-019-00231-w

[2] R. H. Byrne, T. A. Nguyen, D. A. Copp, B. R. Chalamala, and I. Gyuk, "Energy management and optimization methods for grid energy storage systems," *IEEE Access*, vol. 6, pp. 13231–13260, 2018.

[3] "Case 18-E-0130, in the Matter of Energy Storage Deployment Program, Order Establishing Energy Storage Goal and Deployment Policy (issued December 13, 2018)," New York Public Service Commission, 2018.

[4] "Solving Challenges in Energy Storage," US Department of Energy, Office of Technology Transitions. Accessed at www.energy.gov/sites/default/files/2019/07/f64/2018-OTT-Energy-Storage-Spotlight.pdf, 2019.

[5] A. Z. Al Shaqsi, K. Sopian, and A. Al-Hinai, "Review of energy storage services, applications, limitations, and benefits," *Energy Reports*, vol. 6, pp.288–306, 2020.

[6] M. Pasta, C. D. Wessells, R. A. Huggins, and Y. Cui, "A high-rate and long cycle life aqueous electrolyte battery for grid-scale energy storage," *Nature Communications*, vol. 3, 1149, 2012.

[7] T. Bowen, I. Chernyakhovskiy, P. Denholm, and National Renewable Energy Laboratory, "Grid-scale battery storage: frequently asked questions," www.nrel.gov/docs/fy19osti/74426.pdf.

[8] K. Kubota, M. Dahbi, T. Hosaka, S. Kumakura, and S. Komaba, "Towards K-ion and Na-ion batteries as 'beyond Li-ion'," *The Chemical Record*, vol. 18, pp. 459–479, 2018.

[9] H. Chen, T. N. Cong, W. Yang *et al.*, "Progress in electrical energy storage system: a critical review," *Progress in Natural Science*, vol. 19, pp.291–312, 2009.

[10] T. A. Faunce, J. Prest, D. Su, S. J. Hearne, and F. Iacopi, "On-grid batteries for large-scale energy storage: challenges and opportunities for policy and technology," *MRS Energy & Sustainability*, vol. 5, p. E11, 2018.

[11] "DOE OE Global Energy Storage Database," National Technology & Engineering Sciences of Sandia, 2020. Accessed at https://sandia.gov/ess-ssl/gesdb/public/index.html.

[12] J. Hernández, I. Gyuk, and C. Christensen, "DOE global energy storage database – a platform for large scale data analytics and system performance metrics," in *2016 IEEE International Conference on Power System Technology (POWERCON)*, 2016, pp. 1–6. doi: https://doi.org/10.1109/POWERCON.2016.7754009.

[13] "Distributed Energy Resources Database," New York State Energy Research and Development Authority, 2020. Accessed at https://der.nyserda.ny.gov/search/

[14] M. U. Ali, A. Zafar, S. H. Nengroo et al., "Towards a smarter battery management system for electric vehicle applications: a critical review of lithium-ion battery state of charge estimation," *Energies*, vol. 12, p. 446, 2019.

[15] J. O. G. Posada, A. J. R. Rennie, S. P. Villar et al., "Aqueous batteries as grid scale energy storage solutions," *Renewable and Sustainable Energy Reviews*, vol. 68, pp.1174–1182, 2017.

[16] A. S. Subburaj, B. N. Pushpakaran, and S. B. Bayne, "Overview of grid connected renewable energy based battery projects in USA," *Renewable and Sustainable Energy Reviews*, vol. 45, pp.219–234, 2015.

[17] G. J. May, A. Davidson, and B. Monahov, "Lead batteries for utility energy storage: a review," *Journal of Energy Storage*, vol. 15, pp.145–157, 2018.

[18] D. G. Enos, "Chapter 3 – Lead-acid batteries for medium- and large-scale energy storage," in C. Menictas, M. Skyllas-Kazacos, and T. M. Lim, eds., *Advances in Batteries for Medium and Large-Scale Energy Storage*, Woodhead Publishing, 2015, pp. 57–71.

[19] R. Nelson, "The basic chemistry of gas recombination in lead-acid batteries," *JOM*, vol. 53, pp. 28–33, 2001.

[20] F. Beck and P. Rüetschi, "Rechargeable batteries with aqueous electrolytes," *Electrochimica Acta*, vol. 45, pp. 2467–2482, 2000.

[21] A. Cooper, J. Furakawa, L. Lam, and M. Kellaway, "The UltraBattery – a new battery design for a new beginning in hybrid electric vehicle energy storage," *Journal of Power Sources*, vol. 188, pp. 642–649, 2009.

[22] S.-B. Lai, M.-I. Jamesh, X.-C. Wu et al., "A promising energy storage system: rechargeable Ni–Zn battery," *Rare Metals*, vol. 36, pp.381–396, 2017.

[23] V. Pop, "State-of-the-art of battery state-of-charge determination," in V. Pop, H. J. Bergveld, D. Danilov, P. P. L. Regtien, and P. H. L. Notten, eds., *Battery Management Systems: Accurate State-of-Charge Indication for Battery-Powered Applications*, Dordrecht: Springer Netherlands, 2008, pp. 11–45.

[24] F. Putois, "Market for nickel-cadmium batteries," *Journal of Power Sources*, vol. 57, pp.67–70, 1995.

[25] S. K. Dhar, S. R. Ovshinsky, P. R. Gifford *et al.*, "Nickel/metal hydride technology for consumer and electric vehicle batteries – a review and up-date," *Journal of Power Sources*, vol. 65, pp.1–7, 1997.

[26] J. F. Parker, C. N. Chervin, I. R. Pala *et al.*, "Rechargeable nickel–3D zinc batteries: an energy-dense, safer alternative to lithium-ion," *Science*, vol. 356, pp. 415–418, 2017.

[27] B. Dunn, H. Kamath, and J.-M. Tarascon, "Electrical energy storage for the grid: A battery of choices," *Science*, vol. 334, p. 928, 2011.

[28] Z. Yang, J. Zhang, M. C. W. Kintner-Meyer *et al.*, "Electrochemical energy storage for green grid," *Chemical Reviews*, vol. 111, pp. 3577–3613, 2011.

[29] D. H. Doughty, P. C. Butler, A. A. Akhil, N. H. Clark, and J. D. Boyes, "Batteries for large-scale stationary electrical energy storage," *The Electrochemical Society Interface*, vol. 19, pp. 49–53, 2010.

[30] M. Duduta, B. Ho, V. C. Wood *et al.*, "Semi-solid lithium rechargeable flow battery," vol. 1, pp. 511–516, 2011.

[31] X.-Z. Yuan, C. Song, A. Platt *et al.*, "A review of all-vanadium redox flow battery durability: degradation mechanisms and mitigation strategies," vol. 43, pp. 6599–6638, 2019.

[32] K. J. Kim, M.-S. Park, Y.-J. Kim *et al.*, "A technology review of electrodes and reaction mechanisms in vanadium redox flow batteries," *Journal of Materials Chemistry A*, vol. 3, pp. 16913–16933, 2015.

[33] S. Weber, J. F. Peters, M. Baumann, and M. Weil, "Life cycle assessment of a vanadium redox flow battery," *Environmental Science & Technology*, vol. 52, pp. 10864–10873, 2018.

[34] C. Choi, S. Kim, R. Kim *et al.*, "A review of vanadium electrolytes for vanadium redox flow batteries," *Renewable and Sustainable Energy Reviews*, vol. 69, pp.263–274, 2017.

[35] F. Rahman and M. Skyllas-Kazacos, "Solubility of vanadyl sulfate in concentrated sulfuric acid solutions," *Journal of Power Sources*, vol. 72, pp.105–110, 1998.

[36] P. Qian, H. Zhang, J. Chen *et al.*, "A novel electrode-bipolar plate assembly for vanadium redox flow battery applications," *Journal of Power Sources*, vol. 175, pp.613–620, 2008.

[37] G. Kear, A. A. Shah, and F. C. Walsh, "Development of the all-vanadium redox flow battery for energy storage: a review of technological, financial and policy aspects," *International Journal of Energy Research*, vol. 36, pp. 1105–1120, 2012.

[38] W. Xie, R. M. Darling, and M. L. Perry, "Processing and pretreatment effects on vanadium transport in nafion membranes," *Journal of The Electrochemical Society*, vol. 163, pp. A5084–A5089, 2015.

[39] Q. Luo, H. Zhang, J. Chen, P. Qian, and Y. Zhai, "Modification of Nafion membrane using interfacial polymerization for vanadium redox flow battery applications," *Journal of Membrane Science*, vol. 311, pp. 98–103, 2008.

[40] X. Teng, Y. Zhao, J. Xi *et al.*, "Nafion/organic silica modified TiO_2 composite membrane for vanadium redox flow battery via in situ sol–gel reactions," *Journal of Membrane Science*, vol. 341, pp.149–154, 2009.

[41] M. Skyllas-Kazacos, M. H. Chakrabarti, S. A. Hajimolana, F. S. Mjalli, and M. Saleem, "Progress in flow battery research and development," *Journal of The Electrochemical Society*, vol. 158, p. R55, 2011.

[42] F. Pan and Q. Wang, "Redox species of redox flow batteries: a review," *Molecules (Basel, Switzerland)*, vol. 20, pp. 20499–20517, 2015.

[43] H. Zhou, H. Zhang, P. Zhao, and B. Yi, "A comparative study of carbon felt and activated carbon based electrodes for sodium polysulfide/bromine redox flow battery," *Electrochimica Acta*, vol. 51, pp. 6304–6312, 2006.

[44] P. Zhao, H. Zhang, H. Zhou, and B. Yi, "Nickel foam and carbon felt applications for sodium polysulfide/bromine redox flow battery electrodes," *Electrochimica Acta*, vol. 51, pp. 1091–1098, 2005.

[45] P. Leung, X. Li, C. Ponce de León *et al.*, "Progress in redox flow batteries, remaining challenges and their applications in energy storage," *RSC Advances*, vol. 2, pp. 10125–10156, 2012.

[46] C. L. Dean Frankel, S. Minnihan, K. See, and L. Xie, "Flow battery cost reduction: exploring strategies to improve market adoption," *Lux Research State of the Market Report*, https://members.luxresearchinc.com/research/report/15909, 2014.

[47] V. F. K. Mongird, V. Viswanathan, V. Koritarov *et al.*, "Energy storage technology and cost characterization report," https://energystorage.pnnl.gov/pdf/PNNL-28866.pdf, 2019.

[48] D. Reed, E. Thomsen, B. Li *et al.*, "Stack developments in a kW class all vanadium mixed acid redox flow battery at the Pacific Northwest National Laboratory," *Journal of The Electrochemical Society*, vol. 163, pp. A5211–A5219, 2015.

[49] S.-P. Guo, J.-C. Li, Q.-T. Xu, Z. Ma, and H.-G. Xue, "Recent achievements on polyanion-type compounds for sodium-ion batteries: syntheses, crystal chemistry and electrochemical performance," *Journal of Power Sources*, vol. 361, pp.285–299, 2017.

[50] K. Harrington, "Flow battery makers battle over new electrolyte," https://www.aiche.org/chenected/2014/09/flow-battery-makers-battle-over-new-electrolyte, 2014.

[51] A. A. Shinkle, A. E. S. Sleightholme, L. T. Thompson, and C. W. Monroe, "Electrode kinetics in non-aqueous vanadium acetylacetonate redox flow batteries," *Journal of Applied Electrochemistry*, vol. 41, pp. 1191–1199, 2011.

[52] Q. Liu, A. A. Shinkle, Y. Li *et al.*, "Non-aqueous chromium acetylacetonate electrolyte for redox flow batteries," *Electrochemistry Communications*, vol. 12, pp. 1634–1637, 2010.

[53] A. E. S. Sleightholme, A. A. Shinkle, Q. Liu *et al.*, "Non-aqueous manganese acetylacetonate electrolyte for redox flow batteries," *Journal of Power Sources*, vol. 196, pp. 5742–5745, 2011.

[54] M. Duduta, B. Ho, V. C. Wood *et al.*, "Semi-solid lithium rechargeable flow battery," *Advanced Energy Materials*, vol. 1, pp. 511–516, 2011.

[55] Z. Li, S. Li, S. Liu *et al.*, "Electrochemical properties of an all-organic redox flow battery using 2,2,6,6-tetramethyl-1-piperidinyloxy and N-methylphthalimide," *Electrochemical and Solid-State Letters*, vol. 14, p. A171, 2011.

[56] B. Yang, L. Hoober-Burkhardt, F. Wang, G. K. Surya Prakash, and S. R. Narayanan, "An inexpensive aqueous flow battery for large-scale electrical energy storage based on water-soluble organic redox couples," *Journal of The Electrochemical Society*, vol. 161, pp. A1371–A1380, 2014.

[57] L. Wang, A. Abraham, D. M. Lutz *et al.*, "Toward environmentally friendly lithium sulfur batteries: probing the role of electrode design in MoS_2-containing Li–S batteries with a green electrolyte," *ACS Sustainable Chemistry & Engineering*, vol. 7, pp. 5209–5222, 2019.

[58] Y.-X. Wang, B. Zhang, W. Lai *et al.*, "Room-temperature sodium-sulfur batteries: a comprehensive review on research progress and cell chemistry," *Advanced Energy Materials*, vol. 7, p. 1602829, 2017.

[59] J. T. Kummer and N. Weber, "A sodium-sulfur secondary battery," SAE Technical Paper 670179, https://doi.org/10.4271/670179, 1967.

[60] G. Nikiforidis, M. C. M. van de Sanden, and M. N. Tsampas, "High and intermediate temperature sodium–sulfur batteries for energy storage: development, challenges and perspectives," *RSC Advances*, vol. 9, pp. 5649–5673, 2019.

[61] S. Xin, Y.-X. Yin, Y.-G. Guo, and L.-J. Wan, "A high-energy room-temperature sodium-sulfur battery," *Advanced Materials*, vol. 26, pp. 1261–1265, 2014.

[62] R. Steudel and Y. Steudel, "Polysulfide chemistry in sodium–sulfur batteries and related systems – a computational study by G3X(MP2) and PCM calculations," *Chemistry – A European Journal*, vol. 19, pp. 3162–3176, 2013.

[63] X. Yu and A. Manthiram, "Sodium-sulfur batteries with a polymer-coated NASICON-type sodium-ion solid electrolyte," *Matter*, vol. 1, pp.439–451, 2019.

[64] R. D. Armstrong, T. Dickinson, and M. Reid, "Alternating current impedance measurements of the vitreous carbon/sodium polysulphide interphase at 350°C," *Electrochimica Acta*, vol. 21, pp. 935–942, 1976.

[65] G. Kim, Y.-C. Park, Y. Lee *et al.*, "The effect of cathode felt geometries on electrochemical characteristics of sodium sulfur (NaS) cells: planar vs. tubular," *Journal of Power Sources*, vol. 325, pp.238–245, 2016.

[66] R. Okuyama, H. Nakashima, T. Sano, and E. Nomura, "The effect of metal sulfides in the cathode on Na/S battery performance," *Journal of Power Sources*, vol. 93, pp.50–54, 2001.

[67] J. L. Sudworth, "The sodium/sulphur battery," *Journal of Power Sources*, vol. 11, pp. 143–154, 1984.

[68] X. Lu, G. Xia, J. P. Lemmon, and Z. Yang, "Advanced materials for sodium-beta alumina batteries: status, challenges and perspectives," *Journal of Power Sources*, vol. 195, pp. 2431–2442, 2010.

[69] M. M. T. Miyoshi, Y. Kusakabe, H. Hatou *et al.*, US Pat., Application No. 10/246703, 2003.

[70] F. Li, Z. Wei, A. Manthiram *et al.*, "Sodium-based batteries: from critical materials to battery systems," *Journal of Materials Chemistry A*, vol. 7, pp. 9406–9431, 2019.

[71] A. Manthiram and X. Yu, "Ambient temperature sodium–sulfur batteries," *Small*, vol. 11, pp. 2108–2114, 2015.

[72] Y. Ohki, *IEEE Electrical Insulation Magazine*, vol. 33, pp. 59–61, 2017.

[73] X. Tan, Q. Li, and H. Wang, "Advances and trends of energy storage technology in Microgrid," *International Journal of Electrical Power & Energy Systems*, vol. 44, pp. 179–191, 2013.

[74] M. Andriollo, R. Benato, S. Dambone Sessa *et al.*, "Energy intensive electrochemical storage in Italy: 34.8 MW sodium–sulphur secondary cells," *Journal of Energy Storage*, vol. 5, pp.146–155, 2016.

[75] I. Staffell and M. Rustomji, "Maximising the value of electricity storage," *Journal of Energy Storage*, vol. 8, pp.212–225, 2016.

[76] K. B. Hueso, M. Armand, and T. Rojo, "High temperature sodium batteries: status, challenges and future trends," *Energy & Environmental Science*, vol. 6, pp. 734–749, 2013.

[77] J. Sudworth, "The sodium/nickel chloride (ZEBRA) battery," *Journal of Power Sources*, vol. 100, pp. 149–163, 2001.

[78] J. Coetzer, "A new high energy density battery system," *Journal of Power Sources*, vol. 18, pp.377–380, 1986.

[79] X. Gao, Y. Hu, Y. Li *et al*., "High-rate and long-life intermediate-temperature Na–NiCl$_2$ battery with dual-functional Ni–carbon composite nanofiber network," *ACS Applied Materials & Interfaces*, vol. 12, pp. 24767–24776, 2020.

[80] Y. Li, X. Wu, J. Wang *et al*., "Ni-less cathode with 3D free-standing conductive network for planar Na-NiCl2 batteries," *Chemical Engineering Journal*, vol. 387, p. 124059, 2020.

[81] B.-M. Ahn, C.-W. Ahn, B.-D. Hahn *et al*., "Easy approach to realize low cost and high cell capacity in sodium nickel-iron chloride battery," *Composites Part B: Engineering*, vol. 168, pp.442–447, 2019.

[82] X. Zhan, M. E. Bowden, X. Lu *et al*., "A low-cost durable Na-FeCl$_2$ battery with ultrahigh rate capability," *Advanced Energy Materials*, vol. 10, 1903472, 2020.

[83] X. Lu, H. J. Chang, J. F. Bonnett *et al*., "An intermediate-temperature high-performance Na–ZnCl$_2$ battery," *ACS Omega*, vol. 3, pp. 15702–15708, 2018.

[84] X. Lu, G. Li, J. Y. Kim *et al*., "A novel low-cost sodium–zinc chloride battery," *Energy & Environmental Science*, vol. 6, pp. 1837–1843, 2013.

[85] K. B. Hueso, V. Palomares, M. Armand, and T. Rojo, "Challenges and perspectives on high and intermediate-temperature sodium batteries," *Nano Research*, vol. 10, pp. 4082–4114, 2017.

[86] R. Christensen, "Na-NiCl$_2$ batteries." In *Technology Data for Energy storage: November 2018* (pp. 147–160), vol. 183. Danish Energy Agency. https://ens.dk/en/our-services/projections-and-models/technology-data, 2018.

[87] F. Li, Z. X. Wei, A. Manthiram *et al*., "Sodium-based batteries: from critical materials to battery systems," *Journal of Materials Chemistry A*, vol. 7, pp. 9406–9431, Apr. 2019.

[88] M. Y. Li, Z. J. Du, M. A. Khaleel, and I. Belharouak, "Materials and engineering endeavors towards practical sodium-ion batteries," *Energy Storage Materials*, vol. 25, pp. 520–536, Mar. 2020.

[89] P. Adelhelm, P. Hartmann, C. L. Bender *et al*., "From lithium to sodium: cell chemistry of room temperature sodium-air and sodium-sulfur batteries," *Beilstein Journal of Nanotechnology*, vol. 6, pp. 1016–1055, Apr. 2015.

[90] H. Yu, S. Guo, Y. Zhu, M. Ishida, and H. Zhou, "Novel titanium-based O3-type NaTi0.5Ni0.5O2 as a cathode material for sodium ion batteries," *Chemical Communications*, vol. 50, pp. 457–459, 2014.

[91] I. Zatovsky, "NASICON-type Na3V2(PO4)3," *Acta Crystallographica Section E*, vol. 66, p. i12, 2010.

[92] X. Wu, C. Wu, C. Wei *et al.*, "Highly crystallized $Na_2CoFe(CN)_6$ with suppressed lattice defects as superior cathode material for sodium-ion batteries," *ACS Applied Materials & Interfaces*, vol. 8, pp. 5393–5399, 2016.

[93] C. Matei Ghimbeu, J. Górka, V. Simone *et al.*, "Insights on the Na+ ion storage mechanism in hard carbon: discrimination between the porosity, surface functional groups and defects," *Nano Energy*, vol. 44, pp.327–335, 2018.

[94] A. El Kharbachi, O. Zavorotynska, M. Latroche *et al.*, "Exploits, advances and challenges benefiting beyond Li-ion battery technologies," *Journal of Alloys and Compounds*, vol. 817, p. 153261, 2020.

[95] R. D. Shannon, "Revised effective ionic-radii and systematic studies of interatomic distances in halides and chalcogenides," *Acta Crystallographica Section A*, vol. 32, pp. 751–767, 1976.

[96] D. Su, H. J. Ahn, and G. Wang, "Hydrothermal synthesis of alpha-MnO2 and beta-MnO2 nanorods as high capacity cathode materials for sodium ion batteries," *Journal of Materials Chemistry A*, vol. 1, pp. 4845–4850, 2013.

[97] D. W. Su, H. J. Ahn, and G. X. Wang, "Beta-MnO_2 nanorods with exposed tunnel structures as high-performance cathode materials for sodium-ion batteries," *Npg Asia Materials*, vol. 5, Nov. 2013.

[98] J. Huang, A. S. Poyraz, S.-Y. Lee *et al.*, "Silver-containing α-MnO_2 nanorods: electrochemistry in Na-based battery systems," *ACS Applied Materials & Interfaces*, vol. 9, pp. 4333–4342, 2017.

[99] S.-Y. Lee, L. M. Housel, J. Huang *et al.*, "Inhomogeneous structural evolution of silver-containing Alpha-MnO2 nanorods in sodium-ion batteries investigated by comparative transmission electron microscopy approach," *Journal of Power Sources*, vol. 435, 226779, 2019.

[100] J. Qian, C. Wu, Y. Cao *et al.*, "Prussian blue cathode materials for sodium-ion batteries and other ion batteries," *Advanced Energy Materials*, vol. 8, 2018.

[101] X. Wu, M. Sun, S. Guo *et al.*, "Vacancy-free Prussian blue nanocrystals with high capacity and superior cyclability for aqueous sodium-ion batteries," *ChemNanoMat*, vol. 1, pp. 188–193, 2015.

[102] C. D. Wessells, R. A. Huggins, and Y. Cui, "Copper hexacyanoferrate battery electrodes with long cycle life and high power," *Nature Communications*, vol. 2, 550, 2011.

[103] J. Song, L. Wang, Y. Lu *et al.*, "Removal of interstitial H2O in hexacya-nometallates for a superior cathode of a sodium-ion battery," *Journal of the American Chemical Society*, vol. 137, pp. 2658–2664, 2015.

[104] X. Wu, W. Deng, J. Qian *et al.*, "Single-crystal FeFe(CN)6 nanoparticles: a high capacity and high rate cathode for Na-ion batteries," *Journal of Materials Chemistry A*, vol. 1, pp. 10130–10134, 2013.

[105] Y. Jiang, S. Yu, B. Wang *et al.*, "Prussian Blue@C composite as an ultrahigh-rate and long-life sodium-ion battery cathode," *Advanced Functional Materials*, vol. 26, pp. 5315–5321, 2016.

[106] Y. You, H. R. Yao, S. Xin *et al.*, "Subzero-temperature cathode for a sodium-ion battery," *Advanced Materials*, vol. 28, pp. 7243–7248, 2016.

[107] Y. Fang, X. Y. Yu, and X. W. Lou, "A practical high-energy cathode for sodium-ion batteries based on uniform P2-Na0.7CoO2 Microspheres," *Angewandte Chemie International Edition*, vol. 56, pp. 5801–5805, 2017.

[108] S. Komaba, C. Takei, T. Nakayama, A. Ogata, and N. Yabuuchi, "Electrochemical intercalation activity of layered NaCrO2 vs. LiCrO2," *Electrochemistry Communications*, vol. 12, pp. 355–358, 2010.

[109] K. Kubota, T. Asari, H. Yoshida *et al.*, "Understanding the structural evolution and redox mechanism of a $NaFeO_2$–$NaCoO_2$ solid solution for sodium-ion batteries," *Advanced Functional Materials*, vol. 26, pp. 6047–6059, 2016.

[110] S. Komaba, N. Yabuuchi, T. Nakayama *et al.*, "Study on the reversible electrode reaction of $Na_{1-x}Ni_{0.5}Mn_{0.5}O_2$ for a rechargeable sodium-ion battery," *Inorganic Chemistry*, vol. 51, pp. 6211–6220, 2012.

[111] D. Buchholz, A. Moretti, R. Kloepsch *et al.*, "Toward Na-ion batteries – synthesis and characterization of a novel high capacity Na ion intercal-ation material," *Chemistry of Materials*, vol. 25, pp.142–148, 2013.

[112] L. G. Chagas, D. Buchholz, L. M. Wu, B. Vortmann, and S. Passerini, "Unexpected performance of layered sodium-ion cathode material in ionic liquid-based electrolyte," *Journal of Power Sources*, vol. 247, pp. 377–383, Feb. 2014.

[113] J.-Y. Hwang, S.-T. Myung, J. U. Choi *et al.*, "Resolving the degradation pathways of the O3-type layered oxide cathode surface through the nano-scale aluminum oxide coating for high-energy density sodium-ion

batteries," *Journal of Materials Chemistry A*, vol. 5, pp. 23671–23680, 2017.

[114] H. J. Yu, S. H. Guo, Y. B. Zhu, M. Ishida, and H. S. Zhou, "Novel titanium-based O-3-type $NaTi0.5Ni0.5O2$ as a cathode material for sodium ion batteries," *Chemical Communications*, vol. 50, pp. 457–459, 2014.

[115] H.-R. Yao, P.-F. Wang, Y. Gong *et al.*, "Designing air-stable O_3-type cathode materials by combined structure modulation for Na-ion batteries," *Journal of the American Chemical Society*, vol. 139, pp. 8440–8443, 2017.

[116] P. Vassilaras, S. T. Dacek, H. Kim *et al.*, "Communication – O_3-type layered oxide with a quaternary transition metal composition for Na-ion battery cathodes: $NaTi_{0.25}Fe_{0.25}Co_{0.25}Ni_{0.25}O_2$," *Journal of The Electrochemical Society*, vol. 164, pp.A3484–A3486, 2017.

[117] J. Billaud, R. J. Clément, A. R. Armstrong *et al.*, "β-$NaMnO_2$: a high-performance cathode for sodium-ion batteries," *Journal of the American Chemical Society*, vol. 136, pp. 17243–17248, 2014.

[118] P. Barpanda, G. Oyama, S.-i. Nishimura, S.-C. Chung, and A. Yamada, "A 3.8-V earth-abundant sodium battery electrode," *Nature Communications*, vol. 5, 4358, 2014.

[119] K. Saravanan, C. W. Mason, A. Rudola, K. H. Wong, and P. Balaya, "The first report on excellent cycling stability and superior rate capability of $Na_3V_2(PO_4)_3$ for sodium ion batteries," *Advanced Energy Materials*, vol. 3, pp. 444–450, 2013.

[120] Y.-U. Park, D.-H. Seo, H.-S. Kwon *et al.*, "A new high-energy cathode for a Na-ion battery with ultrahigh stability," *Journal of the American Chemical Society*, vol. 135, pp. 13870–13878, 2013.

[121] M. Bianchini, P. Xiao, Y. Wang, and G. Ceder, "Additional sodium insertion into polyanionic cathodes for higher-energy Na-ion batteries," *Advanced Energy Materials*, vol. 7, 2017.

[122] W. Tang, X. Song, Y. Du *et al.*, "High-performance NaFePO4 formed by aqueous ion-exchange and its mechanism for advanced sodium ion batteries," *Journal of Materials Chemistry A*, vol. 4, pp. 4882–4892, 2016.

[123] Y. Fang, L. Xiao, X. Ai, Y. Cao, and H. Yang, "Hierarchical carbon framework wrapped $Na_3V_2(PO_4)_3$ as a superior high-rate and extended lifespan cathode for sodium-ion batteries," *Advanced Materials*, vol. 27, pp. 5895–5900, 2015.

[124] G. Yang, H. Song, M. Wu, and C. Wang, "Porous NaTi2(PO4)3 nano-cubes: a high-rate nonaqueous sodium anode material with more than 10

000 cycle life," *Journal of Materials Chemistry A*, vol. 3, pp. 18718–18726, 2015.

[125] C. Y. Chen, K. Matsumoto, T. Nohira *et al.*, "Pyrophosphate Na2FeP2O7 as a low-cost and high-performance positive electrode material for sodium secondary batteries utilizing an inorganic ionic liquid," *Journal of Power Sources*, vol. 246, pp. 783–787, Jan. 2014.

[126] S. Jiao, J. Tuo, H. Xie *et al.*, "The electrochemical performance of Cu3 [Fe(CN)6]2 as a cathode material for sodium-ion batteries," *Materials Research Bulletin*, vol. 86, pp. 194–200, 2017.

[127] H.-W. Lee, R. Y. Wang, M. Pasta *et al.*, "Manganese hexacyanomanganate open framework as a high-capacity positive electrode material for sodium-ion batteries," *Nature Communications*, vol. 5, 5280, 2014.

[128] M. Takachi, T. Matsuda, and Y. Moritomo, "Cobalt hexacyanoferrate as cathode material for Na+ secondary battery," *Applied Physics Express*, vol. 6, Feb. 2013.

[129] Y. You, X.-L. Wu, Y.-X. Yin, and Y.-G. Guo, "A zero-strain insertion cathode material of nickel ferricyanide for sodium-ion batteries," *Journal of Materials Chemistry A*, vol. 1, pp. 14061–14065, 2013.

[130] H. Lee, Y.-I. Kim, J.-K. Park, and J. W. Choi, "Sodium zinc hexacyanoferrate with a well-defined open framework as a positive electrode for sodium ion batteries," *Chemical Communications*, vol. 48, pp. 8416–8418, 2012.

[131] S. Komaba, W. Murata, T. Ishikawa *et al.*, "Electrochemical Na insertion and solid electrolyte interphase for hard-carbon electrodes and application to Na-ion batteries," *Advanced Functional Materials*, vol. 21, pp. 3859–3867, 2011.

[132] J. Zhao, L. Zhao, K. Chihara *et al.*, "Electrochemical and thermal properties of hard carbon-type anodes for Na-ion batteries," *Journal of Power Sources*, vol. 244, pp.752–757, 2013.

[133] D. A. Stevens and J. R. Dahn, "High capacity anode materials for rechargeable sodium-ion batteries," *Journal of the Electrochemical Society*, vol. 147, pp. 1271–1273, Apr. 2000.

[134] Y. Qiao, R. Han, Y. Liu *et al.*, "Bio-inspired synthesis of an ordered N/P dual-doped porous carbon and application as an anode for sodium-ion batteries," *Chemistry – A European Journal*, vol. 23, pp. 16051–16058, 2017.

[135] Z. Wang, L. Qie, L. Yuan *et al.*, "Functionalized N-doped interconnected carbon nanofibers as an anode material for sodium-ion storage with excellent performance," *Carbon*, vol. 55, pp.328–334, 2013.

[136] Y. Ma, Q. Guo, M. Yang *et al.*, "Highly doped graphene with multi-dopants for high-capacity and ultrastable sodium-ion batteries," *Energy Storage Materials*, vol. 13, pp.134–141, 2018.

[137] Y. Wen, K. He, Y. Zhu *et al.*, "Expanded graphite as superior anode for sodium-ion batteries," *Nature Communications*, vol. 5, 4033, 2014.

[138] W. Xiao, Q. Sun, J. Liu *et al.*, "Utilizing the full capacity of carbon black as anode for Na-ion batteries via solvent co-intercalation," *Nano Research*, vol. 10, pp. 4378–4387, 2017.

[139] Y. Li, L. Mu, Y.-S. Hu *et al.*, "Pitch-derived amorphous carbon as high performance anode for sodium-ion batteries," *Energy Storage Materials*, vol. 2, pp.139–145, 2016.

[140] N. Sun, H. Liu, and B. Xu, "Facile synthesis of high performance hard carbon anode materials for sodium ion batteries," *Journal of Materials Chemistry A*, vol. 3, pp. 20560–20566, 2015.

[141] J. Y. Hwang, S. T. Myung, and Y. K. Sun, "Recent progress in rechargeable potassium batteries," *Advanced Functional Materials*, vol. 28, Oct. 2018.

[142] S. Komaba, T. Hasegawa, M. Dahbi, and K. Kubota, "Potassium intercalation into graphite to realize high-voltage/high-power potassium-ion batteries and potassium-ion capacitors," *Electrochemistry Communications*, vol. 60, pp. 172–175, 2015.

[143] M. M. Huie, D. C. Bock, E. S. Takeuchi, A. C. Marschilok, and K. J. Takeuchi, "Cathode materials for magnesium and magnesium-ion based batteries," *Coordination Chemistry Reviews*, vol. 287, pp. 15–27, Mar. 2015.

[144] M. Rashad, M. Asif, Y. Wang, Z. He, and I. Ahmed, "Recent advances in electrolytes and cathode materials for magnesium and hybrid-ion batteries," *Energy Storage Materials*, vol. 25, pp.342–375, 2020.

[145] M. L. Mao, T. Gao, S. Y. Hou, and C. S. Wang, "A critical review of cathodes for rechargeable Mg batteries," *Chemical Society Reviews*, vol. 47, pp. 8804–8841, Dec. 2018.

[146] H. D. Yoo, I. Shterenberg, Y. Gofer *et al.*, "Mg rechargeable batteries: an on-going challenge," *Energy & Environmental Science*, vol. 6, pp. 2265–2279, 2013.

[147] Y. Xu, X. Deng, Q. Li *et al.*, "Vanadium oxide pillared by interlayer Mg^{2+} ions and water as ultralong-life cathodes for magnesium-ion batteries," *Chem*, vol. 5, pp.1194–1209, 2019.

[148] Y. D. Kondrashev and A. Zaslavskii, "The structure of the modifications of manganese (IV) oxide," *Izvestiya Akademii Nauk SSSR, Seriya Fizicheskaya*, vol. 15, pp. 179–186, 1951.

[149] W. H. Baur, "Rutile-type compounds. V. Refinement of MnO2 and MgF2," *Acta Crystallographica Section B: Structural Crystallography and Crystal Chemistry*, vol. 32, pp. 2200–2204, 1976.

[150] H. Miura, H. Kudou, J. H. Choi, and Y. Hariya, "The crystal structure of ramsdellite from Pirika Mine," *Journal of the Faculty of Science*, vol. 22, pp. 611–617, 1990.

[151] J. E. Post and D. R. Veblen, "Crystal structure determinations of synthetic sodium, magnesium, and potassium birnessite using TEM and the Rietveld method," *American Mineralogist*, vol. 75, pp. 477–489, 1990.

[152] D. Aurbach, Z. Lu, A. Schechter *et al.*, "Prototype systems for rechargeable magnesium batteries," *Nature*, vol. 407, pp.724–727, 2000.

[153] E. Lancry, E. Levi, Y. Gofer *et al.*, "Leaching chemistry and the performance of the Mo6S8 cathodes in rechargeable Mg batteries," *Chemistry of Materials*, vol. 16, pp. 2832–2838, 2004.

[154] E. Lancry, E. Levi, A. Mitelman, S. Malovany, and D. Aurbach, "Molten salt synthesis (MSS) of Cu2Mo6S8 – new way for large-scale production of Chevrel phases," *Journal of Solid State Chemistry*, vol. 179, pp. 1879–1882, 2006.

[155] P. Novák, R. Imhof, and O. Haas, "Magnesium insertion electrodes for rechargeable nonaqueous batteries – a competitive alternative to lithium?," *Electrochimica Acta*, vol. 45, pp. 351–367, 1999.

[156] D. Imamura, M. Miyayama, M. Hibino, and T. Kudo, "Mg intercalation properties into V2O5 gel/carbon composites under high-rate condition," *Journal of the Electrochemical Society*, vol. 150, pp. A753–A758, Jun 2003.

[157] L. Jiao, H. Yuan, Y. Wang, J. Cao, and Y. Wang, "Mg intercalation properties into open-ended vanadium oxide nanotubes," *Electrochemistry Communications*, vol. 7, pp. 431–436, 2005.

[158] G. Gershinsky, H. D. Yoo, Y. Gofer, and D. Aurbach, "Electrochemical and spectroscopic analysis of Mg2+ intercalation into thin film electrodes of layered oxides: V_2O_5 and MoO_3," *Langmuir*, vol. 29, pp. 10964–10972, 2013.

[159] V. Petkov, P. N. Trikalitis, E. S. Bozin *et al.*, "Structure of V_2O_5 ·nH_2O xerogel solved by the atomic pair distribution function technique," *Journal of the American Chemical Society*, vol. 124, pp. 10157–10162, 2002.

[160] N. Sa, T. L. Kinnibrugh, H. Wang *et al.*, "Structural evolution of reversible Mg insertion into a bilayer structure of V_2O_5·nH_2O xerogel material," *Chemistry of Materials*, vol. 28, pp. 2962–2969, 2016.

[161] G. Sai Gautam, P. Canepa, W. D. Richards, R. Malik, and G. Ceder, "Role of structural H_2O in intercalation electrodes: the case of Mg in nanocrystalline xerogel-V_2O_5," *Nano Letters*, vol. 16, pp. 2426–2431, 2016.

[162] P. Novak, W. Scheifele, F. Joho, and O. Haas, "Electrochemical insertion of magnesium into hydrated vanadium bronzes," *Journal of the Electrochemical Society*, vol. 142, pp. 2544–2550, Aug. 1995.

[163] S. H. Lee, R. A. DiLeo, A. C. Marschilok, K. J. Takeuchi, and E. S. Takeuchi, "Sol gel based synthesis and electrochemistry of magnesium vanadium oxide: a promising cathode material for secondary magnesium ion batteries," *Ecs Electrochemistry Letters*, vol. 3, pp. 87–90, 2014.

[164] X. Deng, Y. Xu, Q. An *et al.*, "Manganese ion pre-intercalated hydrated vanadium oxide as a high-performance cathode for magnesium ion batteries," *Journal of Materials Chemistry A*, vol. 7, pp. 10644–10650, 2019.

[165] X. Hu, D. A. Kitchaev, L. Wu *et al.*, "Revealing and rationalizing the rich polytypism of todorokite MnO_2," *Journal of the American Chemical Society*, vol. 140, pp. 6961–6968, 2018.

[166] N. Kumagai, S. Komaba, H. Sakai, and N. Kumagai, "Preparation of todorokite-type manganese-based oxide and its application as lithium and magnesium rechargeable battery cathode," *Journal of Power Sources*, vol. 97–98, pp. 515–517, 2001.

[167] R. Zhang, X. Yu, K.-W. Nam *et al.*, "α-$MnO2$ as a cathode material for rechargeable Mg batteries," *Electrochemistry Communications*, vol. 23, pp. 110–113, 2012.

[168] S. Rasul, S. Suzuki, S. Yamaguchi, and M. Miyayama, "High capacity positive electrodes for secondary Mg-ion batteries," *Electrochimica Acta*, vol. 82, pp. 243–249, 2012.

[169] T. S. Arthur, R. Zhang, C. Ling *et al.*, "Understanding the electrochemical mechanism of K-αMnO_2 for magnesium battery cathodes," *ACS Applied Materials & Interfaces*, vol. 6, pp. 7004–7008, 2014.

[170] L. Wang, K. Asheim, P. E. Vullum, A. M. Svensson, and F. Vullum-Bruer, "Sponge-like porous manganese(II,III) oxide as a highly efficient cathode material for rechargeable magnesium ion batteries," *Chemistry of Materials*, vol. 28, pp. 6459–6470, 2016.

[171] L. Wang, P. E. Vullum, K. Asheim *et al.*, "High capacity Mg batteries based on surface-controlled electrochemical reactions," *Nano Energy*, vol. 48, pp. 227–237, 2018.

[172] M. Cabello, R. Alcántara, F. Nacimiento *et al.*, "Electrochemical and chemical insertion/deinsertion of magnesium in spinel-type MgMn2O4 and lambda-MnO2 for both aqueous and non-aqueous magnesium-ion batteries," *CrystEngComm*, vol. 17, pp. 8728–8735, 2015.

[173] Z. Feng, X. Chen, L. Qiao *et al.*, "Phase-controlled electrochemical activity of epitaxial Mg-spinel thin films," *ACS Applied Materials & Interfaces*, vol. 7, pp. 28438–28443, 2015.

[174] C. Kim, P. J. Phillips, B. Key *et al.*, "Direct observation of reversible magnesium ion intercalation into a spinel oxide host," *Advanced Materials*, vol. 27, pp. 3377–3384, 2015.

[175] O. Mizrahi, N. Amir, E. Pollak *et al.*, "Electrolyte solutions with a wide electrochemical window for rechargeable magnesium batteries," *Journal of The Electrochemical Society*, vol. 155, p. A103, 2008.

[176] D. Aurbach, G. S. Suresh, E. Levi *et al.*, "Progress in rechargeable magnesium battery technology," *Advanced Materials*, vol. 19, pp. 4260-+, Dec. 2007.

[177] Y. L. Liang, R. J. Feng, S. Q. Yang *et al.*, "Rechargeable Mg batteries with graphene-like MoS_2 cathode and ultrasmall Mg nanoparticle anode," *Advanced Materials*, vol. 23, pp. 640–643, Feb. 2011.

[178] A. L. Lipson, S.-D. Han, S. Kim *et al.*, "Nickel hexacyanoferrate, a versatile intercalation host for divalent ions from nonaqueous electrolytes," *Journal of Power Sources*, vol. 325, pp. 646–652, 2016.

[179] Z. Lu, A. Schechter, M. Moshkovich, and D. Aurbach, "On the electrochemical behavior of magnesium electrodes in polar aprotic electrolyte solutions," *Journal of Electroanalytical Chemistry*, vol. 466, pp. 203–217, 1999.

[180] T. D. Gregory, R. J. Hoffman, and R. C. Winterton, "Nonaqueous electrochemistry of magnesium – applications to energy-storage," *Journal of the Electrochemical Society*, vol. 137, pp. 775–780, Mar. 1990.

[181] D. Aurbach, I. Weissman, Y. Gofer, and E. Levi, "Nonaqueous magnesium electrochemistry and its application in secondary batteries," *Chemical Record*, vol. 3, pp. 61–73, 2003.

[182] D. Aurbach, Y. Gofer, Z. Lu *et al.*, "A short review on the comparison between Li battery systems and rechargeable magnesium battery technology," *Journal of Power Sources*, vol. 97–98, pp. 28–32, 2001.

[183] D. Aurbach, H. Gizbar, A. Schechter *et al.*, "Electrolyte solutions for rechargeable magnesium batteries based on organomagnesium chloroaluminate complexes," *Journal of the Electrochemical Society*, vol. 149, pp. A115–A121, Feb. 2002.

[184] E. G. Nelson, S. I. Brody, J. W. Kampf, and B. M. Bartlett, "A magnesium tetraphenylaluminate battery electrolyte exhibits a wide electrochemical potential window and reduces stainless steel corrosion," *Journal of Materials Chemistry A*, vol. 2, pp. 18194–18198, 2014.

[185] N. Pour, Y. Gofer, D. T. Major, and D. Aurbach, "Structural analysis of electrolyte solutions for rechargeable Mg batteries by stereoscopic means and DFT calculations," *Journal of the American Chemical Society*, vol. 133, pp. 6270–6278, 2011.

[186] T. Liu, Y. Shao, G. Li *et al.*, "A facile approach using $MgCl_2$ to formulate high performance Mg_2+ electrolytes for rechargeable Mg batteries," *Journal of Materials Chemistry A*, vol. 2, pp. 3430–3438, 2014.

[187] C. J. Barile, E. C. Barile, K. R. Zavadil, R. G. Nuzzo, and A. A. Gewirth, "Electrolytic conditioning of a magnesium aluminum chloride complex for reversible magnesium deposition," *The Journal of Physical Chemistry C*, vol. 118, pp. 27623–27630, 2014.

[188] K. A. See, Y.-M. Liu, Y. Ha, C. J. Barile, and A. A. Gewirth, "Effect of concentration on the electrochemistry and speciation of the magnesium aluminum chloride complex electrolyte solution," *ACS Applied Materials & Interfaces*, vol. 9, pp. 35729–35739, 2017.

[189] J. Muldoon, C. B. Bucur, A. G. Oliver *et al.*, "Corrosion of magnesium electrolytes: chlorides – the culprit," *Energy & Environmental Science*, vol. 6, pp. 482–487, 2013.

[190] R. Mohtadi, M. Matsui, T. S. Arthur, and S. J. Hwang, "Magnesium borohydride: from hydrogen storage to magnesium battery," *Angewandte Chemie International Edition*, vol. 51, pp. 9780–9783, 2012.

[191] Y. Shao, T. Liu, G. Li *et al.*, "Coordination chemistry in magnesium battery electrolytes: how ligands affect their performance," *Scientific Reports*, vol. 3, 3130, 2013.

[192] J. Niu, Z. Zhang, and D. Aurbach, "Alloy anode materials for recharge-able Mg ion batteries," *Advanced Energy Materials*, vol. n/a, 2000697, 2020.

[193] F. Liu, T. Wang, X. Liu, and L.-Z. Fan, "Challenges and recent progress on key materials for rechargeable magnesium batteries," *Advanced Energy Materials*, vol. n/a, 2000787, 2020.

[194] Y. Shao, M. Gu, X. Li *et al.*, "Highly reversible Mg insertion in nanos-tructured Bi for Mg ion batteries," *Nano Letters*, vol. 14, pp. 255–260, 2014.

[195] K. V. Kravchyk, L. Piveteau, R. Caputo *et al.*, "Colloidal bismuth nanocrystals as a model anode material for rechargeable Mg-ion

batteries: atomistic and mesoscale insights," *ACS Nano*, vol. 12, pp. 8297–8307, 2018.

[196] T. S. Arthur, N. Singh, and M. Matsui, "Electrodeposited Bi, Sb and Bi1-xSbx alloys as anodes for Mg-ion batteries," *Electrochemistry Communications*, vol. 16, pp. 103–106, 2012.

[197] R. A. DiLeo, Q. Zhang, A. C. Marschilok, K. J. Takeuchi, and E. S. Takeuchi, "Composite anodes for secondary magnesium ion batteries prepared via electrodeposition of nanostructured bismuth on carbon nanotube substrates," *ECS Electrochemistry Letters*, vol. 4, pp. A10–A14, 2014.

[198] R. Attias, M. Salama, B. Hirsch, Y. Goffer, and D. Aurbach, "Anode-electrolyte interfaces in secondary magnesium batteries," *Joule*, vol. 3, pp. 27–52, 2019.

[199] N. Singh, T. S. Arthur, C. Ling, M. Matsui, and F. Mizuno, "A high energy-density tin anode for rechargeable magnesium-ion batteries," *Chemical Communications*, vol. 49, pp. 149–151, 2013.

[200] L. Wang, S. S. Welborn, H. Kumar *et al.*, "High-rate and long cycle-life alloy-type magnesium-ion battery anode enabled through (de)magnesiation-induced near-room-temperature solid–liquid phase transformation," *Advanced Energy Materials*, vol. 9, 1902086, 2019.

[201] Y. Cheng, Y. Shao, L. R. Parent *et al.*, "Interface promoted reversible Mg insertion in nanostructured tin–antimony alloys," *Advanced Materials*, vol. 27, pp. 6598–6605, 2015.

[202] J. Niu, H. Gao, W. Ma *et al.*, "Dual phase enhanced superior electrochemical performance of nanoporous bismuth-tin alloy anodes for magnesium-ion batteries," *Energy Storage Materials*, vol. 14, pp. 351–360, 2018.

[203] J. Niu, K. Yin, H. Gao *et al.*, "Composition- and size-modulated porous bismuth–tin biphase alloys as anodes for advanced magnesium ion batteries," *Nanoscale*, vol. 11, pp. 15279–15288, 2019.

[204] M. Song, J. Niu, K. Yin *et al.*, "Self-supporting, eutectic-like, nanoporous biphase bismuth-tin film for high-performance magnesium storage," *Nano Research*, vol. 12, pp. 801–808, 2019.

[205] D.-T. Nguyen, X. M. Tran, J. Kang, and S.-W. Song, "Magnesium storage performance and surface film formation behavior of tin anode material," *ChemElectroChem*, vol. 3, pp. 1813–1819, 2016.

[206] G. Fang, J. Zhou, A. Pan, and S. Liang, "Recent advances in aqueous zinc-ion batteries," *ACS Energy Letters*, vol. 3, pp. 2480–2501, 2018.

[207] B. Y. Tang, L. T. Shan, S. Q. Liang, and J. Zhou, "Issues and opportunities facing aqueous zinc-ion batteries," *Energy & Environmental Science*, vol. 12, pp. 3288–3304, Nov. 2019.

[208] W. W. Xu and Y. Wang, "Recent progress on zinc-ion rechargeable batteries," *Nano-Micro Letters*, vol. 11, Oct 2019.

[209] C. Li, X. Zhang, W. He, G. Xu, and R. Sun, "Cathode materials for rechargeable zinc-ion batteries: from synthesis to mechanism and applications," *Journal of Power Sources*, vol. 449, 227596, 2020.

[210] X. Y. Liu, J. Yi, K. Wu *et al.*, "Rechargeable Zn-MnO2 batteries: advances, challenges and perspectives," *Nanotechnology*, vol. 31, Mar. 2020.

[211] F. Wan and Z. Niu, "Design strategies for vanadium-based aqueous zinc-ion batteries," *Angewandte Chemie International Edition*, vol. 58, pp. 16358–16367, 2019.

[212] B. Lee, H. R. Lee, H. Kim *et al.*, "Elucidating the intercalation mechanism of zinc ions into α-MnO2 for rechargeable zinc batteries," *Chemical Communications*, vol. 51, pp. 9265–9268, 2015.

[213] B. Lee, C. S. Yoon, H. R. Lee *et al.*, "Electrochemically-induced reversible transition from the tunneled to layered polymorphs of manganese dioxide," *Scientific Reports*, vol. 4, 6066, 2014.

[214] C. Wei, C. Xu, B. Li, H. Du, and F. Kang, "Preparation and characterization of manganese dioxides with nano-sized tunnel structures for zinc ion storage," *Journal of Physics and Chemistry of Solids*, vol. 73, pp. 1487–1491, 2012.

[215] C. J. Xu, B. H. Li, H. D. Du, and F. Y. Kang, "Energetic zinc ion chemistry: the rechargeable zinc ion battery," *Angewandte Chemie-International Edition*, vol. 51, pp. 933–935, 2012.

[216] H. Pan, Y. Shao, P. Yan *et al.*, "Reversible aqueous zinc/manganese oxide energy storage from conversion reactions," *Nature Energy*, vol. 1, 16039, 2016.

[217] W. Sun, F. Wang, S. Hou *et al.*, "Zn/MnO$_2$ battery chemistry with H+ and Zn2+ coinsertion," *Journal of the American Chemical Society*, vol. 139, pp. 9775–9778, 2017.

[218] N. Zhang, F. Cheng, J. Liu *et al.*, "Rechargeable aqueous zinc-manganese dioxide batteries with high energy and power densities," *Nature Communications*, vol. 8, 405, 2017.

[219] S. Islam, M. H. Alfaruqi, V. Mathew *et al.*, "Facile synthesis and the exploration of the zinc storage mechanism of β-MnO2 nanorods with exposed (101) planes as a novel cathode material for high performance eco-friendly zinc-ion batteries," *Journal of Materials Chemistry A*, vol. 5, pp. 23299–23309, 2017.

[220] M. H. Alfaruqi, V. Mathew, J. Gim *et al.*, "Electrochemically induced structural transformation in a γ-MnO$_2$ cathode of a high capacity zinc-ion battery system," *Chemistry of Materials*, vol. 27, pp. 3609–3620, 2015.

[221] J. Lee, J. B. Ju, W. I. Cho, B. W. Cho, and S. H. Oh, "Todorokite-type MnO2 as a zinc-ion intercalating material," *Electrochimica Acta*, vol. 112, pp. 138–143, 2013.

[222] Y. Jin, L. F. Zou, L. L. Liu *et al.*, "Joint charge storage for high-rate aqueous zinc-manganese dioxide batteries," *Advanced Materials*, vol. 31, Jul. 2019.

[223] K. W. Nam, H. Kim, J. H. Choi, and J. W. Choi, "Crystal water for high performance layered manganese oxide cathodes in aqueous rechargeable zinc batteries," *Energy & Environmental Science*, vol. 12, pp. 1999–2009, Jun. 2019.

[224] H. Ren, J. Zhao, L. Yang *et al.*, "Inverse opal manganese dioxide constructed by few-layered ultrathin nanosheets as high-performance cathodes for aqueous zinc-ion batteries," *Nano Research*, vol. 12, pp. 1347–1353, 2019.

[225] J. S. Ko, M. B. Sassin, J. F. Parker, D. R. Rolison, and Jeffrey W. Long, "Combining battery-like and pseudocapacitive charge storage in 3D MnOx@carbon electrode architectures for zinc-ion cells," *Sustainable Energy & Fuels*, vol. 2, pp. 626–636, 2018.

[226] M. H. Alfaruqi, J. Gim, S. Kim *et al.*, "A layered δ-MnO2 nanoflake cathode with high zinc-storage capacities for eco-friendly battery applications," *Electrochemistry Communications*, vol. 60, pp. 121–125, 2015.

[227] N. Zhang, F. Cheng, Y. Liu *et al.*, "Cation-deficient spinel ZnMn$_2$O$_4$ cathode in Zn(CF$_3$SO$_3$)$_2$ electrolyte for rechargeable aqueous Zn-ion battery," *Journal of the American Chemical Society*, vol. 138, pp. 12894–12901, 2016.

[228] G. Fang, C. Zhu, M. Chen *et al.*, "Suppressing manganese dissolution in potassium manganate with rich oxygen defects engaged high-energy-density and durable aqueous zinc-ion battery," *Advanced Functional Materials*, vol. 29, 1808375, 2019.

[229] L. T. Shan, J. Zhou, W. Y. Zhang *et al.*, "Highly reversible phase transition endows V$_6$O$_{13}$ with enhanced performance as aqueous zinc-ion battery cathode," *Energy Technology*, vol. 7, Jun 2019.

[230] J. Zhou, L. T. Shan, Z. X. Wu *et al.*, "Investigation of V2O5 as a low-cost rechargeable aqueous zinc ion battery cathode," *Chemical Communications*, vol. 54, pp. 4457–4460, Apr. 2018.

[231] M. Yan, P. He, Y. Chen *et al.*, "Water-lubricated intercalation in V_2O_5 ·nH_2O for high-capacity and high-rate aqueous rechargeable zinc batteries," *Advanced Materials*, vol. 30, 1703725, 2018.

[232] D. Kundu, B. D. Adams, V. Duffort, S. H. Vajargah, and L. F. Nazar, "A high-capacity and long-life aqueous rechargeable zinc battery using a metal oxide intercalation cathode," *Nature Energy*, vol. 1, 16119, 2016.

[233] C. Xia, J. Guo, P. Li, X. Zhang, and H. N. Alshareef, "Highly stable aqueous zinc-ion storage using a layered calcium vanadium oxide bronze cathode," *Angewandte Chemie International Edition*, vol. 57, pp. 3943–3948, 2018.

[234] P. He, G. Zhang, X. Liao *et al.*, "Sodium ion stabilized vanadium oxide nanowire cathode for high-performance zinc-ion batteries," *Advanced Energy Materials*, vol. 8, 1702463, 2018.

[235] Y. Yang, Y. Tang, G. Fang *et al.*, "Li+ intercalated V2O5·nH2O with enlarged layer spacing and fast ion diffusion as an aqueous zinc-ion battery cathode," *Energy & Environmental Science*, vol. 11, pp. 3157–3162, 2018.

[236] P. He, Y. L. Quan, X. Xu *et al.*, "High-performance aqueous zinc-ion battery based on layered $H_2V_3O_8$ nanowire cathode," *Small*, vol. 13, Dec 2017.

[237] P. Hu, T. Zhu, X. Wang *et al.*, "Highly durable $Na_2V_6O_{16}$·$1.63H_2O$ nanowire cathode for aqueous zinc-ion battery," *Nano Letters*, vol. 18, pp. 1758–1763, 2018.

[238] F. Wan, L. Zhang, X. Dai *et al.*, "Aqueous rechargeable zinc/sodium vanadate batteries with enhanced performance from simultaneous insertion of dual carriers," *Nature Communications*, vol. 9, 1656, 2018.

[239] V. Yufit, F. Tariq, D. S. Eastwood *et al.*, "Operando visualization and multi-scale tomography studies of dendrite formation and dissolution in zinc batteries," *Joule*, vol. 3, pp. 485–502, 2019.

[240] W. J. Lu, C. X. Xie, H. M. Zhang, and X. F. Li, "Inhibition of zinc dendrite growth in zinc-based batteries," *Chemsuschem*, vol. 11, pp. 3996–4006, Dec. 2018.

[241] J. F. Parker, C. N. Chervin, E. S. Nelson, D. R. Rolison, and J. W. Long, "Wiring zinc in three dimensions re-writes battery performance – dendrite-free cycling," *Energy & Environmental Science*, vol. 7, pp. 1117–1124, 2014.

[242] B. J. Hopkins, M. B. Sassin, C. N. Chervin *et al.*, "Fabricating architected zinc electrodes with unprecedented volumetric capacity in rechargeable alkaline cells," *Energy Storage Materials*, vol. 27, pp. 370–376, 2020.

[243] J. F. Parker, C. N. Chervin, I. R. Pala *et al.*, "Rechargeable nickel–3D zinc batteries: an energy-dense, safer alternative to lithium-ion," *Science*, vol. 356, 415, 2017.

[244] Z. Kang, C. Wu, L. Dong *et al.*, "3D porous copper skeleton supported zinc anode toward high capacity and long cycle life zinc ion batteries," *ACS Sustainable Chemistry & Engineering*, vol. 7, pp. 3364–3371, 2019.

[245] H. F. Li, C. J. Xu, C. P. Han *et al.*, "Enhancement on cycle performance of Zn anodes by activated carbon modification for neutral rechargeable zinc ion batteries," *Journal of the Electrochemical Society*, vol. 162, pp. A1439–A1444, 2015.

[246] X. Wang, F. Wang, L. Wang *et al.*, "An aqueous rechargeable Zn//Co$_3$O$_4$ battery with high energy density and good cycling behavior," *Advanced Materials*, vol. 28, pp. 4904–4911, 2016.

[247] L.-P. Wang, N.-W. Li, T.-S. Wang *et al.*, "Conductive graphite fiber as a stable host for zinc metal anodes," *Electrochimica Acta*, vol. 244, pp. 172–177, 2017.

[248] W. Qiu, Y. Li, A. You *et al.*, "High-performance flexible quasi-solid-state Zn–MnO2 battery based on MnO2 nanorod arrays coated 3D porous nitrogen-doped carbon cloth," *Journal of Materials Chemistry A*, vol. 5, pp. 14838–14846, 2017.

[249] Z. Zhao, J. Zhao, Z. Hu *et al.*, "Long-life and deeply rechargeable aqueous Zn anodes enabled by a multifunctional brightener-inspired interphase," *Energy & Environmental Science*, vol. 12, pp. 1938–1949, 2019.

[250] K. N. Zhao, C. X. Wang, Y. H. Yu *et al.*, "Ultrathin surface coating enables stabilized zinc metal anode," *Advanced Materials Interfaces*, vol. 5, 1800848, Aug. 2018.

[251] L. Kang, M. Cui, F. Jiang *et al.*, "Nanoporous CaCO$_3$ coatings enabled uniform Zn stripping/plating for long-Life zinc rechargeable aqueous batteries," *Advanced Energy Materials*, vol. 8, 1801090, 2018.

[252] X. Xie, S. Liang, J. Gao *et al.*, "Manipulating the ion-transfer kinetics and interface stability for high-performance zinc metal anodes," *Energy & Environmental Science*, vol. 13, pp. 503–510, 2020.

[253] J. Ming, J. Guo, C. Xia, W. Wang, and H. N. Alshareef, "Zinc-ion batteries: materials, mechanisms, and applications," *Materials Science and Engineering: R: Reports*, vol. 135, pp. 58–84, 2019.

[254] F. Ding, W. Xu, G. L. Graff *et al.*, "Dendrite-free lithium deposition via self-healing electrostatic shield mechanism," *Journal of the American Chemical Society*, vol. 135, pp. 4450–4456, 2013.

[255] S. J. Banik and R. Akolkar, "Suppressing dendrite growth during zinc electrodeposition by PEG-200 additive," *Journal of the Electrochemical Society*, vol. 160, pp. D519–D523, 2013.

[256] S. J. Banik and R. Akolkar, "Suppressing dendritic growth during alkaline zinc electrodeposition using polyethylenimine additive," *Electrochimica Acta*, vol. 179, pp. 475–481, 2015.

[257] Z. Hou, X. Zhang, X. Li *et al.*, "Surfactant widens the electrochemical window of an aqueous electrolyte for better rechargeable aqueous sodium/zinc battery," *Journal of Materials Chemistry A*, vol. 5, pp. 730–738, 2017.

[258] C. W. Lee, K. Sathiyanarayanan, S. W. Eom, H. S. Kim, and M. S. Yun, "Novel electrochemical behavior of zinc anodes in zinc/air batteries in the presence of additives," *Journal of Power Sources*, vol. 159, pp. 1474–1477, 2006.

[259] M. H. Alfaruqi, S. Islam, D. Y. Putro *et al.*, "Structural transformation and electrochemical study of layered MnO2 in rechargeable aqueous zinc-ion battery," *Electrochimica Acta*, vol. 276, pp. 1–11, 2018.

[260] X. W. Wu, Y. H. Xiang, Q. J. Peng *et al.*, "Green-low-cost rechargeable aqueous zinc-ion batteries using hollow porous spinel ZnMn2O4 as the cathode material," *Journal of Materials Chemistry A*, vol. 5, pp. 17990–17997, Sep. 2017.

[261] S. H. Kim and S. M. Oh, "Degradation mechanism of layered MnO2 cathodes in Zn/ZnSO4/MnO2 rechargeable cells," *Journal of Power Sources*, vol. 72, pp. 150–158, 1998.

[262] J. H. Jo, Y.-K. Sun, and S.-T. Myung, "Hollandite-type Al-doped VO1.52 (OH)0.77 as a zinc ion insertion host material," *Journal of Materials Chemistry A*, vol. 5, pp. 8367–8375, 2017.

[263] J. Lai, H. Zhu, X. Zhu, H. Koritala, and Y. Wang, "Interlayer-expanded $V_6O_{13} \cdot nH_2O$ architecture constructed for an advanced rechargeable aqueous zinc-ion battery," *ACS Applied Energy Materials*, vol. 2, pp. 1988–1996, 2019.

[264] Y. Li and J. Lu, "Metal–air batteries: will they be the future electrochemical energy storage device of choice?," *ACS Energy Letters*, vol. 2, pp. 1370–1377, 2017.

[265] J. Zhang, Q. Zhou, Y. Tang, L. Zhang, and Y. Li, "Zinc–air batteries: are they ready for prime time?," *Chemical Science*, vol. 10, pp. 8924–8929, 2019.

[266] Y. Li and H. Dai, "Recent advances in zinc–air batteries," *Chemical Society Reviews*, vol. 43, pp. 5257–5275, 2014.

[267] N.-T. Suen, S.-F. Hung, Q. Quan *et al.*, "Electrocatalysis for the oxygen evolution reaction: recent development and future perspectives," *Chemical Society Reviews*, vol. 46, pp. 337–365, 2017.

[268] A. R. Mainar, L. C. Colmenares, O. Leonet *et al.*, "Manganese oxide catalysts for secondary zinc air batteries: from electrocatalytic activity to bifunctional air electrode performance," *Electrochimica Acta*, vol. 217, pp. 80–91, 2016.

[269] D. Stock, S. Dongmo, J. Janek, and D. Schröder, "Benchmarking anode concepts: the future of electrically rechargeable zinc–air batteries," *ACS Energy Letters*, vol. 4, pp. 1287–1300, 2019.

[270] J. F. Parker, J. S. Ko, D. R. Rolison, and J. W. Long, "Translating materials-level performance into device-relevant metrics for zinc-based batteries," *Joule*, vol. 2, pp. 2519–2527, 2018.

[271] A. R. Mainar, E. Iruin, L. C. Colmenares *et al.*, "An overview of progress in electrolytes for secondary zinc-air batteries and other storage systems based on zinc," *Journal of Energy Storage*, vol. 15, pp. 304–328, 2018.

[272] S. Clark, A. R. Mainar, E. Iruin *et al.*, "Towards rechargeable zinc–air batteries with aqueous chloride electrolytes," *Journal of Materials Chemistry A*, vol. 7, pp. 11387–11399, 2019.

[273] J. Weaver, "Zinc-air battery being deployed in New York aims for extremely low $45/kWh cost," *PV Magazine*, Jan. 27, 2020.

[274] J. Fu, Z. P. Cano, M. G. Park *et al.*, "*Electrically rechargeable zinc–air batteries: progress, challenges, and perspectives*," vol. 29, 1604685, 2017.

[275] X. Fan, B. Liu, J. Liu *et al.*, "Battery technologies for grid-level large-scale electrical energy storage," *Transactions of Tianjin University*, vol. 26, pp. 92–103, 2020.

Acknowledgments

ACM acknowledges support from the Department of Energy, Office of Electricity, administered through Sandia National Laboratories, Purchase Order #1955692. GPW and LW acknowledge support from Brookhaven National Laboratory.

Cambridge Elements $\equiv$

Grid Energy Storage

Babu Chalamala

Sandia National Laboratories

Dr. Babu Chalamala is manager of the Energy Storage Technologies and Systems Department at Sandia National Laboratories. He received his Ph.D. degree in Physics from the University of North Texas and has extensive corporate and startup experience spanning several years. He is an IEEE Fellow and chair of the IEEE Energy Storage and Stationary Battery Committee.

Vincent Sprenkle

Pacific Northwest National Laboratory

Dr. Vincent Sprenkle is chief scientist at Pacific Northwest National Laboratory (PNNL), and program manager for the Department of Energy's Electricity Energy Storage Program at PNNL. His work focusses on electrochemical energy storage technologies to enable renewable integration and improve grid support. He has a Ph.D. from the University of Missouri and holds 25 patents on fuel cells and batteries.

Imre Gyuk

US Department of Energy

Dr. Imre Gyuk is Director of the Energy Storage Research Program at DOE's Office of Electricity. For the past 2 decades, he has directed work on a wide portfolio of storage technologies for a broad spectrum of applications. He has a Ph.D. from Purdue University. His work has won prestigious awards including 12 R&D 100 Awards, the Phil Symons Award from ESA, and a Lifetime Achievement Award from NAATBatt.

Ralph D. Masiello

Quanta Technology

Dr. Ralph D. Masiello is a senior advisor at Quanta Technology, and developed smart grid roadmaps for several US independent system operators and the California Energy Commission. With a Ph.D. from MIT in electrical engineering, he is a Life Fellow of the IEEE, member of the US National Academy of Engineering, and won the 2009 IEEE Power Engineering Concordia Award.

Raymond Byrne

Sandia National Laboratories

Dr. Raymond Byrne is the manager of the Power Electronics and Energy Conversion Systems Department at Sandia National Laboratories, and works on optimal control of energy storage to maximize grid integration of renewables. He is a Fellow of the IEEE and recipient of the IEEE Millennium Medal.

About the Series

This new Elements series is perfect for practicing engineers who need to incorporate grid energy storage into their electricity infrastructure and seek comprehensive technical details about all aspects of grid energy storage. The addressed topics will span from energy storage materials to the engineering of energy storage systems.

Cumulatively, the Elements series will cover energy storage technologies, distributed energy storage systems, power electronics and control systems for grid and off-grid storage, the application of stationary energy storage systems for improving grid stability and reliability, and the integration of energy storage in electricity infrastructure. This series is co-published in collaboration with the Materials Research Society.

MATERIALS RESEARCH SOCIETY®
Advancing materials. Improving the quality of life.

Cambridge Elements $\equiv$

Grid Energy Storage

Elements in the Series

Beyond Li-ion Batteries for Grid-Scale Energy Storage
Garrett P. Wheeler, Lei Wang, and Amy C. Marschilok

A full series listing is available at: www.cambridge.org/EGES

Printed in the United States
by Baker & Taylor Publisher Services